Practical Guide to Applied Conformal Prediction in Python

Learn and apply the best uncertainty frameworks to your industry applications

Valery Manokhin

‹packt›

BIRMINGHAM—MUMBAI

Practical Guide to Applied Conformal Prediction in Python

Group Product Manager: Niranjan Naikwadi

Publishing Product Manager: Nitin Nainani

Content Development Editor: Priyanka Soam

Technical Editor: Devanshi Ayare

Copy Editor: Safis Editing

Project Coordinator: Shambhavi Mishra

Proofreader: Safis Editing

Indexer: Rekha Nair

Production Designer: Prafulla Nikalje

Marketing Coordinator: Vinishka Kalra

First published: Nov 2023

Production reference: 2041223

Published by Packt Publishing Ltd.
Grosvenor House
11 St Paul's Square
Birmingham
B3 1RB, UK.

ISBN 978-1-80512-276-0

www.packtpub.com

Foreword

In statistical and machine learning, it is rare to encounter a technique that blends deep mathematical rigor with practical simplicity. Conformal Prediction is one such gem. Rooted in solid probability theory, it transcends academic theory to find wide-ranging applications in the real world.

Valery, studied under the inventor of Conformal Prediction, compiles in this book a treasure trove of practical knowledge, tailored for practicing data scientists. His work makes Conformal Prediction not only accessible but intuitively understandable, bridging the gap between complex theory and practical application.

This book stands out for its unique approach to demystifying Conformal Prediction. It eschews the often esoteric and dense theoretical exposition common in statistical texts, opting instead for clarity and comprehensibility. This approach makes the powerful techniques of Conformal Prediction accessible to a broader range of machine learning practitioners.

The applications of Conformal Prediction are vast and varied, and this book delves into them with meticulous detail. From classification and regression to time series analysis to computer vision, and language models. Each application is explored thoroughly with examples to provide practitioners with practical guidance on applying these methods in their work.

This book will be an essential reference for machine learning engineers and data scientists who seek to incorporate **uncertainty quantification** (**UQ**) to models that they develop and deploy, a critical element that has been missing in machine learning. UQ is critical to understand prediction reliability, providing safety during model deployment and potential model weakness identification during model development and testing.

For example, **Python Interpretable Machine Learning** (**PiML**) toolkit that my team developed and applied in banking incorporating Conformal Prediction to identify regions of inputs where models are less reliable (higher prediction uncertainties).

Agus Sudjianto, PhD

Executive Vice President

Head of Corporate Model Risk Wells Fargo

Contributors

About the author

Valery Manokhin, is the leading expert in the field of machine learning and Conformal Prediction. He holds a Ph.D.in Machine Learning from Royal Holloway, University of London. His doctoral work was supervised by the creator of Conformal Prediction, Vladimir Vovk, and focused on developing new methods for quantifying uncertainty in machine learning models.

Valery has published extensively in leading machine learning journals, and his Ph.D. dissertation 'Machine Learning for Probabilistic Prediction' is read by thousands of people across the world. He is also the creator of "Awesome Conformal Prediction," the most popular resource and GitHub repository for all things Conformal Prediction.

About the reviewers

Eleftherios (Lefteris) Koulierakis is a senior data scientist with a diverse international working background. He holds an Engineering Doctorate in data science from Eindhoven University of Technology. He has demonstrated a consistent track record of innovation, notably as the lead inventor of several machine learning patents primarily applicable to the semiconductor industry. He has architected and developed numerous machine learning and deep learning solutions for anomaly detection, image processing, forecasting, and predictive maintenance applications. Embracing collaborations, he has experience in guiding data science teams toward successful product deliveries and also has experience in supporting team members to reach their full potential.

Rahul Vishwakarma is a Senior Member of IEEE and has worked in the industry for over twelve years. During his tenure at Dell Technologies, he drove solutions for Data Protection and assisted customers in safeguarding data with Data Domain. During his position as a Solution Architect at **Hewlett Packard Enterprise (HPE)**, he designed reference architectures for the Converged System for SAP HANA. He holds more than 65 US Patents in machine learning, data storage, persistent memory, DNA storage, and blockchain domains. His current research interests include addressing bias, explainability, and uncertainty quantification of machine learning models.

Table of Contents

Part 1: Introduction

1

2

Part 2: Conformal Prediction Framework

3

4

5

Part 3: Applications of Conformal Prediction

6

7

10

Conformal Prediction for Natural Language Processing   169

Part 4: Advanced Topics

11

Handling Imbalanced Data   181

12

Multi-Class Conformal Prediction   193

Index   205

Other Books You May Enjoy   214

Preface

Embark on an insightful journey with "Practical Guide to Applied Conformal Prediction in Python," your comprehensive guide to mastering uncertainty quantification in machine learning. This book unfolds the complexities of Conformal Prediction, focusing on practical applications that span classification, regression, forecasting, computer vision, and natural language processing. It also delves into sophisticated techniques for addressing imbalanced datasets and multi-class classification challenges, presenting case studies that bridge theory with real-world practice.

This resource is meticulously crafted for a diverse readership, including data scientists, machine learning engineers, industry professionals, researchers, academics, and students interested in mastering uncertainty quantification and conformal prediction within their respective fields.

Whether you're starting your journey in data science or looking to deepen your existing expertise, this book provides the foundational knowledge and advanced strategies necessary to navigate uncertainty quantification in machine learning confidently.

With "Practical Guide to Applied Conformal Prediction in Python," you gain more than knowledge; you gain the power to apply cutting-edge techniques to industry applications, enhancing the precision and reliability of your predictive models. Embrace this opportunity to elevate your career in machine learning by harnessing the potential of Conformal Prediction.

Who is this book for?

This publication is a must-read for those fascinated by Conformal Prediction, catering to a broad spectrum of professionals and learners. It is specifically designed for data scientists, machine learning engineers, educators and scholars, research professionals, application developers, students with a zest for data, analytical experts, and statisticians dedicated to expanding their knowledge.

What this book covers

Chapter 1, Introducing Conformal Prediction, The opening chapter of *Practical Guide to Applied Conformal Prediction in Python* serves as a fundamental introduction to the book's core theme—Conformal Prediction. It lays the foundation by elucidating Conformal Prediction's purpose as a robust framework for effectively quantifying prediction uncertainty and enhancing trust in machine learning models.

Within this chapter, we embark on a journey through the historical evolution and the burgeoning acclaim of this transformative framework. Key concepts and principles that underpin Conformal Prediction are explored, shedding light on its manifold advantages. The chapter underscores how Conformal Prediction stands apart from conventional machine learning techniques. It achieves this distinction by furnishing prediction regions and confidence measures, all underpinned by finite sample validity guarantees, all while eschewing the need for restrictive distributional assumptions.

Chapter 2, Overview of Conformal Prediction, In the second chapter of *Practical Guide to Applied Conformal Prediction in Python*, we embark on a comprehensive journey into the realm of Conformal Prediction, focusing on its pivotal role in quantifying prediction uncertainty.

This chapter commences by addressing the crucial need for quantifying uncertainty in predictions and introduces the concepts of aleatoric and epistemic uncertainty. It emphasizes the distinct advantages offered by Conformal Prediction in comparison to conventional statistical, Bayesian, and fuzzy logic methods. These advantages include the assurance of coverage, freedom from distributional constraints, and compatibility with a wide array of machine learning models.

A significant portion of the chapter is devoted to elucidating how Conformal Prediction operates in a classification context. It unveils the intricate process of using nonconformity scores to gauge the alignment between predictions and the training data distribution. These scores are then transformed into p-values and confidence levels, forming the foundation for constructing prediction sets.

Chapter 2 provides readers with a deep understanding of Conformal Prediction's principles and its profound significance in quantifying uncertainty. This knowledge proves particularly invaluable in critical applications where dependable confidence estimates must accompany predictions, enhancing the overall trustworthiness of the outcomes.

Chapter 3, Fundamentals of Conformal Prediction, dives into the fundamentals and mathematical foundations underlying Conformal Prediction. It explains the basic components like nonconformity measures, calibration sets, and the prediction process.

It covers different types of nonconformity measures for classification and regression, explaining their strengths and weaknesses. Popular choices like hinge loss, margin, and normalized error are discussed.

The chapter illustrates how to compute nonconformity scores, p-values, confidence levels, and credibility levels with examples. It also explains the role of calibration sets, online vs offline conformal prediction, and unconditional vs conditional coverage.

Overall, *Chapter 3* equips readers with a strong grasp of the core concepts and mathematical workings of Conformal Prediction. By mastering these foundations, practitioners can apply Conformal Prediction to enhance the reliability of predictions across various machine learning tasks.

Chapter 4, Validity and Efficiency of Conformal Prediction, extends the concepts introduced in the previous chapter and delves into the crucial notions of validity and efficiency. Through practical examples, readers will discover the importance of accurate (unbiased) prediction models.

This chapter will delve into the definitions, metrics, and real-world instances of valid and efficient models. We'll also explore the inherent validity guarantees offered by conformal prediction. By the chapter's conclusion, you'll possess the knowledge needed to evaluate and enhance the validity and efficiency of your predictive models, opening doors to more dependable and impactful applications in your respective fields.

Chapter 5, Types of Conformal Predictors, This chapter explores the intriguing realm of conformal predictors, exploring their various types and unique attributes. Key topics covered encompass the foundational principles of conformal prediction and its relevance in machine learning. The chapter explains both classical transductive and inductive conformal predictors, guiding readers in selecting the right type of conformal predictor tailored to their specific problem needs. Additionally, practical use cases of conformal predictors in binary classification, multiclass classification, and regression are also presented.

Chapter 6, Conformal Prediction for Classification, This chapter explores different types of conformal predictors for quantifying uncertainty in machine learning predictions. It covers the foundations of classical **Transductive Conformal Prediction** (**TCP**) and the more efficient **Inductive Conformal Prediction** (**ICP**). TCP leverages the full dataset for training but requires model retraining for each new prediction. ICP splits data into training and calibration sets, achieving computational speedup by training once. Tradeoffs between the variants are discussed.

The chapter provides algorithmic descriptions for applying TCP and ICP to classification and regression problems. It steps through calculating nonconformity scores, p-values, and prediction regions in detail using code examples.

Guidelines are given on choosing the right conformal predictor based on factors like data size, real-time requirements, and computational constraints. Example use cases illustrate when TCP or ICP would be preferable.

We also introduce specialized techniques within conformal prediction called Venn-ABERS predictors.

Overall, the chapter offers readers a solid grasp of the different types of conformal predictors available and how to select the optimal approach based on the problem context.

Chapter 7, Conformal Prediction for Regression, This chapter provides a comprehensive guide to uncertainty quantification for regression problems using Conformal Prediction. It covers the need for uncertainty quantification, techniques for generating prediction intervals, Conformal Prediction frameworks tailored for regression, and advanced methods like Conformalized Quantile Regression, Jackknife+ and Conformal Predictive Distributions. Readers will learn the theory and practical application of Conformal Prediction for producing valid, calibrated prediction intervals and distributions. The chapter includes detailed explanations and code illustrations using real-world housing price data and Python libraries to give hands-on experience applying these methods. Overall, readers will gain the knowledge to reliably quantify uncertainty and construct well-calibrated prediction regions for regression problems.

Chapter 8, Conformal Prediction for Time Series and Forecasting, This chapter is dedicated to the application of Conformal Prediction in the realm of time series forecasting.

The chapter initiates with an exploration of the significance of uncertainty quantification in forecasting, emphasizing the concept of prediction intervals. It covers diverse approaches for generating prediction intervals, encompassing parametric methods, non-parametric techniques like bootstrapping, Bayesian approaches, and Conformal Prediction.

Practical implementations of Conformal Prediction for time series are showcased using libraries such as Amazon Fortuna (EnbPI method), Nixtla (statsforecast package), and NeuralProphet. Code examples are provided, illustrating the generation of prediction intervals and the evaluation of validity.

In essence, Chapter 8 equips readers with practical tools and techniques to leverage Conformal Prediction for creating reliable and well-calibrated prediction intervals in time series forecasting models. By incorporating these methods, forecasters can effectively quantify uncertainty and bolster the robustness of their forecasts.

Chapter 9, Conformal Prediction for Computer Vision, In this chapter, we embark on a journey through the application of Conformal Prediction in the realm of computer vision.

The chapter commences by underscoring the paramount significance of uncertainty quantification in vision tasks, particularly in domains with safety-critical implications like medical imaging and autonomous driving. It addresses a common challenge in modern deep learning—overconfident and miscalibrated predictions.

Diverse uncertainty quantification methods are explored before highlighting the unique advantages of Conformal Prediction, including its distribution-free guarantees.

Practical applications of Conformal Prediction in image classification are vividly demonstrated, with a focus on the RAPS algorithm, renowned for generating compact and stable prediction sets. The chapter provides code examples illustrating the construction of classifiers with well-calibrated prediction sets on ImageNet data, employing various Conformal Prediction approaches.

In essence, Chapter 9 equips readers with an understanding of the value of uncertainty quantification in computer vision systems. It offers hands-on experience in harnessing Conformal Prediction to craft dependable image classifiers complete with valid confidence estimates.

Chapter 10, Conformal Prediction for Natural Language Processing, This chapter ventures into the realm of uncertainty quantification in Natural Language Processing (NLP), leveraging the power of Conformal Prediction.

The chapter commences by delving into the inherent ambiguity that characterizes language and the repercussions of miscalibrated predictions stemming from intricate deep learning models.

Various approaches to uncertainty quantification, such as Bayesian methods, bootstrapping, and out-of-distribution detection, are thoughtfully compared. The mechanics of applying conformal prediction to NLP are demystified, encompassing the computation of nonconformity scores and p-values.

The advantages of adopting conformal prediction for NLP are eloquently outlined, including distribution-free guarantees, interpretability, and adaptivity. The chapter also delves into contemporary research, highlighting how conformal prediction enhances reliability, safety, and trust in large language models.

Chapter 11, Handling Imbalanced Data, This chapter explores solutions for the common machine learning challenge of imbalanced data, where one class heavily outweighs others. It explains why this skewed distribution poses complex problems for predictive modeling.

The chapter compares various traditional approaches like oversampling and SMOTE, noting their pitfalls regarding poor model calibration. It then introduces Conformal Prediction as an innovative method to handle imbalanced data without compromising reliability.

Through code examples on a real-world credit card fraud detection dataset, the chapter demonstrates applying conformal prediction for probability calibration even with highly skewed data. Readers will learn best practices for tackling imbalance issues while ensuring decision-ready probabilistic forecasting.

Chapter 12, Multi-Class Conformal Prediction, This final chapter explores multi-class classification and how conformal prediction can be applied to problems with more than two outcome categories. It covers evaluation metrics like precision, recall, F1 score, log loss, and Brier score for assessing model performance.

The chapter explains techniques to extend binary classification algorithms like support vector machines or neural networks to multi-class contexts using one-vs-all and one-vs-one strategies.

It then demonstrates how conformal prediction can provide prediction sets or intervals for each class with validity guarantees. Advanced methods like Venn-ABERS predictors for multi-class probability estimation are also introduced.

Through code examples, the chapter shows how to implement inductive conformal prediction on multi-class problems, outputting predictions with credibility and confidence scores. Readers will learn best practices for applying conformal prediction to classification tasks with multiple potential classes.

To get the most out of this book

You will need a working Python environment on your computer. We recommend using Python 3.6 or later.

Ensure that you have essential libraries, such as scikit-learn, NumPy, and Matplotlib, installed. If not, you can easily install them using Conda or pip.

The notebooks can be run both locally or by using Google Colab (`https://colab.research.google.com`).

Software/hardware covered in the book	Operating system requirements
Python	Windows, macOS, or Linux
Colab (to run notebooks in Google Cloud)	Windows, macOS, or Linux
MAPIE	Windows, macOS, or Linux
Amazon Fortuna	Windows, macOS, or Linux
NIxtla statsforecast	Windows, macOS, or Linux
NeuralProphet	Windows, macOS, or Linux

If you are using the digital version of this book, we advise you to access the code from the book's GitHub repository (a link is available in the next section). Doing so will help you avoid any potential errors related to the copying and pasting of code.

Download the example code files

You can download the example code files for this book from GitHub at `https://github.com/PacktPublishing/Practical-Guide-to-Applied-Conformal-Prediction`. If there's an update to the code, it will be updated in the GitHub repository.

We also have other code bundles from our rich catalog of books and videos available at `https://github.com/PacktPublishing/`. Check them out!

Conventions used

There are a number of text conventions used throughout this book.

`Code in text`: Indicates code words in text, database table names, folder names, filenames, file extensions, pathnames, dummy URLs, user input, and Twitter handles. Here is an example: First, we must create ICP classifiers by using a wrapper from `nonconformist`

A block of code is set as follows:

```
y_pred_calib = model.predict(X_calib)
y_pred_score_calib = model.predict_proba(X_calib)

y_pred_test = model.predict(X_test)
y_pred_score_test = model.predict_proba(X_test)
```

> **Tips or important notes**
> Appear like this.

Get in touch

Feedback from our readers is always welcome.

General feedback: If you have questions about any aspect of this book, email us at customercare@packtpub.com and mention the book title in the subject of your message.

Errata: Although we have taken every care to ensure the accuracy of our content, mistakes do happen. If you have found a mistake in this book, we would be grateful if you would report this to us. Please visit www.packtpub.com/support/errata and fill in the form.

Piracy: If you come across any illegal copies of our works in any form on the internet, we would be grateful if you would provide us with the location address or website name. Please contact us at copyright@packt.com with a link to the material.

If you are interested in becoming an author: If there is a topic that you have expertise in and you are interested in either writing or contributing to a book, please visit authors.packtpub.com.

Share Your Thoughts

Once you've read *Practical Guide to Applied Conformal Prediction in Python*, we'd love to hear your thoughts! Please click here to go straight to the Amazon review page for this book and share your feedback.

Your review is important to us and the tech community and will help us make sure we're delivering excellent quality content.

Download a free PDF copy of this book

Thanks for purchasing this book!

Do you like to read on the go but are unable to carry your print books everywhere? Is your eBook purchase not compatible with the device of your choice?

Don't worry, now with every Packt book you get a DRM-free PDF version of that book at no cost.

Read anywhere, any place, on any device. Search, copy, and paste code from your favorite technical books directly into your application.

The perks don't stop there, you can get exclusive access to discounts, newsletters, and great free content in your inbox daily

Follow these simple steps to get the benefits:

1. Scan the QR code or visit the link below

https://packt.link/free-ebook/9781805122760

2. Submit your proof of purchase
3. That's it! We'll send your free PDF and other benefits to your email directly

Part 1: Introduction

This part will introduce you to conformal prediction. It will explain in detail the type of problem that conformal prediction can address and outline the general ideas on which it is based.

This part has the following chapters:

- *Chapter 1, Introducing Conformal Prediction*
- *Chapter 2, Overview of Conformal Prediction*

1

Introducing Conformal Prediction

This book is about conformal prediction, a modern framework for uncertainty quantification that is becoming increasingly popular in industry and academia.

Machine learning and AI applications are everywhere. In the realm of machine learning, prediction is a fundamental task. Given a training dataset, we train a machine learning model to make predictions on new data.

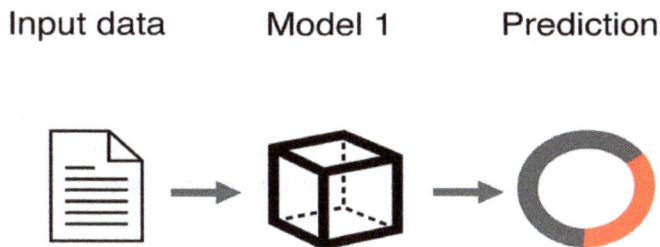

Figure 1.1 – Machine learning prediction model

However, in many real-world applications, the predictions made by statistical, machine learning, and deep learning models are often incorrect or unreliable because of various factors, such as insufficient or incomplete data, issues arising during the modeling process, or simply because of the randomness and complexities of the underlying problem.

Predictions made by machine learning models often come without the uncertainty quantification required for confident and reliable decision-making. This is where conformal prediction comes in. By providing a clear measure of the reliability of its predictions, conformal prediction enhances the trustworthiness and explainability of machine learning models, making them more transparent and user-friendly for decision-makers.

This chapter will introduce conformal prediction and explore how it can be applied in practical settings.

In this chapter, we're going to cover the following main topics:

- Introduction to conformal prediction
- The origins of conformal prediction
- How conformal prediction differs from traditional machine learning
- The p-value and its role in conformal prediction

The chapter will provide a practical understanding of conformal prediction and its applications. By the end of this chapter, you will be able to understand how conformal prediction can be applied to your own machine learning models to improve their reliability and interpretability.

Technical requirements

This book uses Python. The code for this book is hosted on GitHub and can be found here: `https://github.com/PacktPublishing/Practical-Guide-to-Applied-Conformal-Prediction` You can run notebooks locally or upload them to Google Colab (`https://colab.research.google.com/`).

Introduction to conformal prediction

In this section, we will introduce conformal prediction and explain how it can be used to improve the reliability of predictions produced by statistical, machine learning, and deep learning models. We will provide an overview of the key ideas and concepts behind conformal prediction, including its underlying principles and benefits. By the end of this section, you will have a solid understanding of conformal prediction and why it is an important framework to know.

Conformal prediction is a powerful machine learning framework that provides valid confidence measures for individual predictions. This means that when you make a prediction using any model from the conformal prediction framework, you can also measure your confidence in that prediction.

This is incredibly useful in many practical applications where it is crucial to have reliable and interpretable predictions. For example, in medical diagnosis, conformal prediction can provide a confidence level that a tumor is malignant versus benign. This enables physicians to make more informed treatment decisions based on the prediction confidence. In finance, conformal prediction can provide prediction intervals estimating financial risk. This allows investors to quantify upside and downside risks.

Specifically, conformal prediction can determine a 95% chance a tumor is malignant, giving physicians high confidence in a cancer diagnosis. Or, it can predict an 80% probability that a stock price will fall between $50 and $60 next month, providing an estimated trading range. Conformal prediction increases trust and is valuable in real-world applications such as medical diagnosis and financial forecasting by delivering quantifiable confidence in predictions.

The key benefit of conformal prediction is that it provides valid confidence measures for individual predictions. A conformal prediction model usually provides a prediction in the form of a prediction interval or prediction set with a specified confidence level, for example, 95%. In classification problems, conformal prediction can also calibrate class probabilities, enhancing confidence and informed decision-making.

In conformal prediction, "coverage" denotes the likelihood that the predicted region – whether a set of potential outcomes in classification tasks or a prediction interval in regression tasks – accurately encompasses the true values. Essentially, if you choose a coverage of 95%, it means there's a 95% chance that the true values fall within the provided prediction set or interval.

We call such prediction regions "valid." The requirement for the validity of predictions is crucial to ensure that the model does not contain prediction bias and is especially important in consequential applications such as health, finance, and self-driving cars. Valid predictions are a prerequisite of trust in the machine learning model that has produced this prediction.

While there are alternative approaches to uncertainty quantification, such as Bayesian methods, Monte Carlo methods, and bootstrapping, to provide validity guarantees, such approaches require distribution-specific assumptions about the data – for example, an assumption that the data follows a normal distribution. However, the true underlying distribution of real-world data is generally unknown. Conversely, conformal prediction does not make distributional assumptions and can provide validity guarantees without making assumptions about the specifics of data distribution. This makes conformal prediction more broadly applicable to real data that may not satisfy common statistical assumptions such as normality, smoothness, and so on.

In practice, the need for distribution-specific assumptions limits the ability of methods such as Bayesian inference or bootstrapping to make formally rigorous statements about arbitrary real data sources. There is no guarantee that predictions from such methods will have the claimed confidence level or coverage across all data types, since the assumptions may not hold. This can create a mismatch between the confidence level communicated to users and the actual coverage achieved, leading to inaccurate decisions and misleading users about the reliability of model predictions.

Conformal prediction sidesteps these issues by providing distribution-free finite sample validity guarantees without relying on hard-to-verify distributional assumptions about the data. This makes conformal prediction confidence estimates more trustworthy and robust for real-world applications.

Conformal prediction has multiple benefits:

- **Guaranteed coverage**: Conformal prediction guarantees the validity of prediction regions automatically. Any method from the conformal prediction framework guarantees the validity of prediction regions by mathematical design. In comparison, alternative methods output predictions that do not provide any validity guarantees. By way of an example, the popular NGBoost package does not produce valid prediction intervals (you can read more about it at the following link: `https://medium.com/@valeman/does-ngboost-work-evaluating-ngboost-against-critical-criteria-for-good-probabilistic-prediction-28c4871c1bab`).

- **Distribution-free**: Conformal prediction is distribution-free and can be applied to any data distribution regardless of the properties of the distribution as long as the data is exchangeable. Exchangeability means that the order or index of the data points does not matter – shuffling or permuting the data points will not change the overall data distribution. For example, exchangeability assumes that observation 1, 2, 3 has the same distribution as observation 2, 3, 1 or 3, 1, 2. This is a weaker assumption than IID and is required to provide validity guarantees. Unlike many classical statistical models, conformal prediction does not make assumptions such as the data following a normal distribution. The data can have any distribution, even with irregularities such as fat tails. The only requirement is exchangeability. By relying only on exchangeability rather than strict distributional assumptions, conformal prediction provides finite sample guarantees on prediction coverage that are distribution-free and applicable to any data source.

- **Model-agnostic**: Conformal prediction can be applied to any prediction model that produces point predictions in classification, regression, time series, computer vision, NLP, reinforcement learning, or other statistical, machine learning, and deep learning tasks. Conformal prediction has been successfully applied to many innovative model types, including recent innovations such as diffusion models and **large language models (LLMs)**. Conformal prediction does not require the model to be statistical, machine learning, or deep learning. It could be any model of any type, for example, a business heuristic developed by domain experts. If you have a model to make point predictions, you can use conformal prediction as an uncertainty quantification layer on top of your point prediction model to obtain a well-calibrated, reliable, and safe probabilistic prediction model.

- **Non-intrusive**: Conformal prediction stands out in its simplicity and efficiency. Rather than overhauling your existing point prediction model, it seamlessly integrates with it. For businesses with established models already in production, this is a game changer. And for data scientists, the process is even more exciting. Simply overlay your point prediction model with the uncertainty quantification layer provided by conformal prediction, and you're equipped with a state-of-the-art probabilistic prediction model.

- **Dataset size**: Conformal prediction stands apart from typical statistical methods that depend on stringent data distribution assumptions, such as normality, or need vast datasets for solid guarantees. It offers inherent mathematical assurances of valid predictions without bias, irrespective of the dataset's size. While smaller datasets may yield broader prediction intervals in regression tasks (or larger sets in classification), conformal prediction remains consistently valid. The validity is assured no matter the dataset size, underlying prediction model, or data distribution, making it a unique and unmatched method for uncertainty quantification.

- **Easy to use**: A few years back, the adoption of conformal prediction was limited due to the scarcity of open source libraries, even though esteemed universities and major corporations such as Microsoft had been utilizing it for years. Fast forward to today, and the landscape has dramatically shifted. There's a rich selection of top-tier Python packages such as MAPIE and Amazon Fortuna, among others. This means that generating well-calibrated probabilistic predictions via conformal prediction is just a few lines of code away, making it straightforward to integrate into business applications. Furthermore, platforms such as KNIME have democratized its use, offering conformal prediction through low-code or no-code solutions.

- **Fast**: The most widely embraced conformal prediction variant, inductive conformal prediction, stands out because it operates efficiently without the need to retrain the foundational model. In contrast, other methods, such as Bayesian networks, often necessitate retraining. This distinction means that inductive conformal prediction offers a streamlined approach, eliminating the time and computational costs associated with repeated model retraining.

- **Non-intrusive**: Unlike many uncertainty quantification techniques, conformal prediction seamlessly integrates without altering the underlying point prediction models. Its non-invasive nature is cost-effective and convenient, especially compared to other methods that demand potentially costly and complex adjustments to the machine or deep learning models. The benefits of using conformal prediction are truly incredible. You might be interested to know how conformal prediction achieves the unique and powerful benefits that it offers to its users.

The key objective of conformal prediction is to provide valid confidence measures that adapt based on the difficulty of making individual predictions. Conformal prediction uses "nonconformity measures" to assess how well new observations fit with previous observations.

The overall workflow consists of the following steps:

1. A conformal predictor learns from past training examples to quantify uncertainty around predictions for the new observations.

2. When quantifying uncertainty around predictions for the new observations, it calculates nonconformity scores, measuring how different or "nonconforming" the new observation is, compared to the training set (in the classical transductive version of conformal prediction) or calibration (in the most popular variant of conformal prediction – inductive conformal prediction).

3. These nonconformity scores are used to determine whether the new observation falls within the range of values expected based on the training data.

4. The model calculates personalized confidence measures and prediction sets (in classification problems) or prediction intervals (in regression problems and time series forecasting) for each prediction.

The magic of conformal prediction lies in these nonconformity measures, which allow the model to evaluate each new prediction in the context of the previously seen data. This simple but powerful approach results in finite sample coverage guarantees adapted to the intrinsic difficulty of making a given prediction. The validity holds for any data distribution, prediction algorithm, or dataset size.

In this book, we will talk interchangeably about nonconformity and conformity measures; one is the inverse of the other, and depending on the application, it will be more convenient to use either conformity or nonconformity measures.

Understanding conformity measures

A conformity measure is a critical component of conformal prediction and is essentially a function that assigns a numerical score (conformity score) to each object in a dataset. The conformity score indicates how well a new observation fits the observed data. When making a new prediction, we can use the conformity measure to calculate a conformity score for the new observation and compare it to the conformity scores of the previous observations. Based on this comparison, we can calculate a measure of confidence for our prediction. The conformity score indicates a degree of confidence in the prediction.

The choice of conformity measure is a key step in conformal prediction. The conformity measure determines how we assess how similar new observations are to past examples. There are many options for defining conformity measures depending on the problem.

In a classification setting, a simple conformity measure could calculate the probability scores assigned to each class by the prediction model for a new observation. The class with the highest probability would have the best conformity or match to the training data.

The key advantage of conformal prediction is that we obtain valid prediction regions regardless of the conformity measure used. This is because conformal prediction relies only on the order induced by the conformity measure rather than its exact form.

So, we have the flexibility to incorporate domain knowledge in designing an appropriate conformity measure for the problem at hand. If the measure ranks how well new observations match past data, conformal prediction can be used to deliver finite sample coverage guarantees.

While all conformal predictors provide valid prediction regions, the choice of conformity measure impacts their efficiency. Efficiency relates to the width of the prediction intervals or sets – narrower intervals contain more valuable information for decision-making.

Though validity holds for any conformity measure, thoughtfully choosing one tailored to the application can improve efficiency and produce narrower, more useful prediction intervals. The intervals should also be adaptable based on the model's uncertainty – expanding for difficult predictions and contracting for clear ones.

Let's illustrate this with an example. Say we have a dataset of patients diagnosed with a disease, with features such as age, gender, and test results. We want to predict whether new patients are at risk.

A simple conformity measure could calculate how similar the feature values are between new patients and those in the training data. New patients very different from the data would get low conformity scores and wide prediction intervals, indicating high uncertainty. While this conformity measure would produce valid intervals, we can improve efficiency with a more tailored approach.

By carefully selecting conformity measures aligned to our prediction problem and domain knowledge, we can obtain high-quality conformal predictors that provide both validity and efficiency.

We will now talk briefly about the origins of conformal prediction.

The origins of conformal prediction

The origins of conformal prediction are documented in *Gentle Introduction to Conformal Prediction and Distribution-Free Uncertainty Quantification* by Anastasios N. Angelopoulos and Stephen Bates (https://arxiv.org/abs/2107.07511).

> **Note**
>
> Conformal prediction was invented by my PhD supervisor Prof. Vladimir Vovk, a professor at Royal Holloway University of London. Vladimir Vovk graduated from Moscow State University, where he studied mathematics and became a student of one of the most notable mathematicians of the 20th century, Andrey Kolmogorov. During this time, initial ideas that later gave rise to the invention of conformal prediction appeared.
>
> The first edition of *Algorithmic Learning in a Random World* (https://link.springer.com/book/10.1007/b106715) by Vladimir Vovk, Alexander Gammerman, and Glenn Shafer was published in 2005. The second edition of the book was published in 2022 (https://link.springer.com/book/10.1007/978-3-031-06649-8).

Conformal prediction was popularized in United States academia by Professor Larry Wasserman (Carnegie Mellon) and his collaborators, who have published some key papers and introduced conformal prediction to many other researchers in the United States.

> **Note**
>
> In 2022, I finished my PhD in machine learning. In the same year, I created *Awesome Conformal Prediction* (https://github.com/valeman/awesome-conformal-prediction) – the most comprehensive professionally curated resource on conformal prediction, which has since received thousands of GitHub stars.

Conformal prediction has grown rapidly from a niche research area into a mainstream framework for uncertainty quantification. The field has exploded in recent years, with over 1,000 research papers on conformal prediction estimated to be published in 2023 alone.

This surge of research reflects the increasing popularity and applicability of conformal prediction across academia and industry. Major technology companies such as Microsoft, Amazon, DeepMind, and NVIDIA now conduct research into and apply conformal prediction. The framework has also been adopted in high-stakes domains such as healthcare and finance, where validity and reliability are critical.

In just over two decades since its introduction, conformal prediction has cemented itself as one of the premier and most trusted approaches for uncertainty quantification in machine learning. The field will continue to expand as more practitioners recognize the value of conformal prediction's finite sample guarantees compared to traditional statistical methods reliant on asymptotic theory and unverifiable distributional assumptions. With growing research and adoption, conformal prediction is poised to become a standard tool for any application requiring rigorous uncertainty estimates alongside point predictions.

At *NeurIPS 2022*, one of the prominent mathematicians of our time, Emmanuel Candes (Stanford), delivered a key invited talk titled *Conformal Prediction in 2022* (`https://slideslive.com/38996063/conformal-prediction-in-2022?ref=speaker-43789`) to tens of thousands of attendees. In his talk, Emmanuel Candes said:

> *Conformal inference methods are becoming all the rage in academia and industry alike. In a nutshell, these methods deliver exact prediction intervals for future observations without making any distributional assumption whatsoever other than having IID, and more generally, exchangeable data.*

The future of conformal prediction

For years, I have promoted conformal prediction as the premier framework for reliable probabilistic predictions. Excitingly, over the last 2-3 years, there has been an explosion of interest in and adoption of conformal prediction, including by major tech leaders such as Amazon, Microsoft, Google, and DeepMind. Many universities and companies are researching conformal prediction, actively developing real-world applications, and releasing open source libraries such as MAPIE and Amazon Fortuna.

These trends will only accelerate as more practitioners recognize the power of conformal prediction for trustworthy uncertainty quantification. As renowned machine learning researcher Michael I. Jordan noted at the ICML 2021 workshop (`https://icml.cc/virtual/2021/workshop/8373`): "*Conformal prediction ideas are THE answer to UQ (uncertainty quantification); I think it's the best I have seen – it's simple, generalizable, and so on.*"

Conformal prediction has an incredibly bright future as an indispensable tool for quantifying uncertainty in machine learning. Several key reasons are driving the momentum and adoption of conformal prediction:

- **Simplicity**: Conformal prediction is straightforward to understand and implement, making it accessible to practitioners without deep statistical expertise. At its core is the intuitive idea of measuring how new observations conform to past data.
- **Flexibility**: It can be applied to any machine learning model and data distribution. No modifications to existing predictors are needed. This model-agnostic property greatly expands the applicability of conformal prediction.
- **Theoretical guarantees**: The finite sample coverage guarantees provide an unmatched level of reliability compared to traditional statistical methods reliant on asymptotic theory.

These advantages perfectly position conformal prediction as the gold standard for uncertainty quantification in machine learning applications where trustworthy confidence estimates are critical.

I am confident that the adoption of conformal prediction in academic research and industry will accelerate rapidly in the coming years. Its simple yet powerful approach is cementing its place as an essential tool for any practitioner or organization managing predictive uncertainty. The next few years will be incredibly exciting as we realize the full potential of this game-changing framework.

How conformal prediction differs from traditional machine learning

Conformal prediction allows the production of well-calibrated probabilistic predictions for any statistical, machine learning, or deep learning model. This is achieved without relying on restrictive assumptions required by other methods such as Bayesian techniques, Monte Carlo simulation, and bootstrapping. Importantly, conformal prediction does not require subjective priors. It provides mathematically guaranteed, well-calibrated predictions every time – regardless of the underlying prediction model, data distribution, or dataset size.

A key limitation of traditional machine learning is the need for more reasonable confidence measures for individual predictions. Models may have excellent overall performance but not be able to quantify uncertainty for a given input reliably.

Conformal prediction solves this by outputting prediction regions and confidence measures with statistical validity guarantees. It achieves this distribution-free reliability without needing to modify the underlying predictor.

While machine learning and deep learning models struggle to quantify uncertainty, limiting user trust, conformal prediction has been successfully applied to consequential real-world problems. Examples include diagnosing depression, drug discovery, and predicting risks of cancer and stroke.

By delivering trustworthy individual prediction uncertainties, conformal prediction unlocks the potential of machine learning for high-stakes applications requiring confidence measures alongside predictions.

Conformal prediction has invaluable applications across many high-stakes domains where reliable confidence estimates are critical:

- **Medicine**: Conformal prediction can improve trust in AI severity ratings for medical imaging, assisting radiologists in disease grading using MRI scans
- **Health**: It can detect anomalies and provide reliable anomaly scores to inform treatment decisions
- **Self-driving cars**: Conformal prediction can improve autonomous vehicle safety by providing reliable prediction intervals for pedestrian positions and trajectories
- **Finance**: It can ensure algorithmic fairness in lending by providing well-calibrated predictions
- **Recommender systems**: Augmenting recommenders with conformal prediction can improve recommendations by guaranteeing high-quality suggestions and minimizing erroneous items

By delivering rigorous confidence estimates, conformal prediction unlocks reliable and ethical AI applications in medicine, transportation, finance, and beyond. Its validity guarantees make it invaluable for high-stakes decisions relying on machine learning.

The p-value and its role in conformal prediction

In conformal prediction, p-values are key in constructing prediction regions and intervals with a guaranteed confidence level. However, their purpose is different than in traditional statistical hypothesis testing.

Let's walk through an example binary classification task to understand how this works. Suppose we want to predict whether a patient has a medical condition based on their symptoms and characteristics:

1. First, we calculate a nonconformity score that measures how different or "nonconforming" the new patient is compared to previously seen patients. We can define this score in various ways, such as the distance between feature values.

2. Next, we temporarily assign the patient each possible label – 0 (no condition) and 1 (has condition) – and recalculate the nonconformity score with that assigned label.

If the score is similar to scores for past patients with label 0, then label 0 conforms well to the data. To measure this fit statistically, we compute the p-value by comparing the test object's "strangeness" using the test object's nonconformity score to the nonconformity scores of the previous patients.

If this p-value exceeds our chosen significance level, we add label 0 to the prediction set since it fits the data. We repeat this for label 1 to see whether we should include it.

Ultimately, we construct a prediction set containing all labels whose nonconformity scores result in p-values exceeding the significance level. This provides finite sample guarantees on the confidence level of the prediction regions.

So, in summary, nonconformity scores measure how well each potential label conforms to the training data. Then p-values let us convert these scores into statistically rigorous prediction sets and intervals. The two concepts work hand in hand within the conformal prediction framework.

Summary

In this chapter, we have introduced conformal prediction and explained the multiple benefits of this powerful framework for reliably quantifying the uncertainty of predictions to improve trust in machine learning models.

We explained that the key benefit of conformal prediction is that, unlike any other probabilistic prediction framework, it provides valid probabilistic predictions accompanied by confidence measures, regardless of the underlying model, the dataset size, and the data distribution.

We then explored the origins of conformal prediction and saw how it has recently become a very popular framework adopted by leading universities and companies.

Finally, we looked at how conformal prediction differs from traditional machine learning and learned about the role of p-values in conformal prediction.

In *Chapter 2*, we will explain why conformal prediction is a valuable tool for quantifying the uncertainty of predictions, especially in critical settings such as healthcare, self-driving cars, and finance. We will also discuss the concept of uncertainty quantification and how the conformal prediction framework has successfully addressed the challenge of quantifying uncertainty.

2

Overview of Conformal Prediction

In today's world, where data plays a crucial role in decision making, it has become increasingly important to measure uncertainty in predictions. To achieve this, a relatively new framework called **conformal prediction** is gaining popularity. This framework provides probabilistic predictions that are not only robust and reliable but also trustworthy. It is a powerful tool that offers measures of confidence, accuracy, and reliability for a given prediction, allowing users to make informed choices with more certainty.

This chapter will provide an overview of conformal prediction.

It will explain why conformal prediction is a valuable tool for quantifying the uncertainty of predictions, especially in critical settings such as healthcare, self-driving cars, and finance. We will also discuss the concept of **uncertainty quantification** (**UQ**) and how the conformal prediction framework has successfully addressed the challenge of quantifying uncertainty. By the end of this chapter, you will have a better understanding of conformal prediction and its potential applications in various fields.

In this chapter, we'll cover the following topics:

- Understanding uncertainty quantification
- Different ways to quantify uncertainty
- Quantifying uncertainty using conformal prediction

Understanding uncertainty quantification

Uncertainty is an inherent part of any prediction, as there are always factors that are unknown or difficult to measure. Predictions are typically made based on incomplete data or models that are unable to capture the full complexity of the real world. As a result, the predictions are subject to various sources of uncertainty, including randomness, bias, and model errors.

To mitigate the risks associated with uncertainty, it is essential to quantify it accurately. By quantifying uncertainty, we can estimate the range of possible outcomes and assess the degree of confidence we can have in our predictions. This information can be used to make informed decisions and to identify areas where further research or data collection is needed.

UQ is a field of study that helps us measure how much we don't know when we make predictions. UQ tries to estimate the probability of outcomes even if some aspects of the system under study are not known exactly.

Uncertainty is often broken down into two types: **aleatoric** and **epistemic**. Aleatoric uncertainty is caused by the inherent randomness and unpredictability of the system being studied, while epistemic uncertainty arises from our lack of knowledge about the system.

Aleatoric uncertainty

Aleatoric uncertainty refers to the type of uncertainty that is caused by inherent randomness and unpredictability in a system. Here are a few examples of aleatoric uncertainty:

- **Rolling a dice**: The outcome of rolling a dice is inherently random and unpredictable. You cannot predict with certainty what number will come up on the dice, so there is aleatoric uncertainty associated with this process.

- **Weather forecasting**: Weather is a complex system that is influenced by many factors, some of which are inherently random and difficult to predict. For example, the exact path and strength of a storm system may be difficult to predict due to aleatoric uncertainty.

- **Stock market fluctuations**: The stock market is a complex system that is influenced by many factors, including human behavior, economic conditions, and news events. Some of these factors are inherently unpredictable and may cause fluctuations in stock prices, leading to aleatoric uncertainty.

- **Traffic flow**: Traffic flow is a complex system that is influenced by many factors, including the number of vehicles on the road, road conditions, and driver behavior. The exact flow of traffic at any given time is difficult to predict with certainty, leading to aleatoric uncertainty.

Let's cover epistemic uncertainty next.

Epistemic uncertainty

Epistemic uncertainty refers to the type of uncertainty that arises from a lack of knowledge or understanding about a system. Here are a few examples of epistemic uncertainty:

- **Medical diagnosis**: Diagnosing a medical condition involves making predictions based on available data, such as symptoms, medical history, and test results. However, there may be epistemic uncertainty associated with the diagnosis if there is incomplete or unreliable data, or if the medical condition is not well understood.

- **Financial forecasting**: Predicting future economic conditions is a complex task that involves making predictions based on a wide range of factors, including interest rates, inflation, and political events. However, there is always epistemic uncertainty associated with these predictions due to our limited understanding of the economic system.

- **Natural disaster prediction**: Predicting the likelihood and severity of natural disasters, such as earthquakes or hurricanes, involves making predictions based on data from past events and current environmental conditions. However, there may be epistemic uncertainty associated with these predictions if we do not have a complete understanding of the underlying physical processes involved.

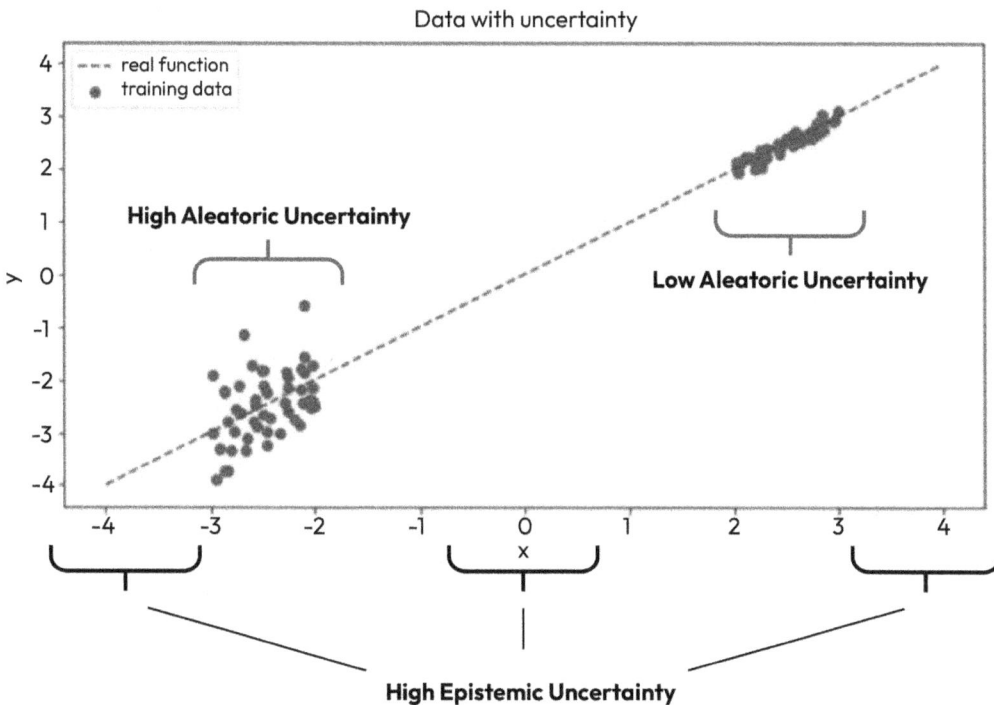

Figure 2.1 – Illustration of aleatoric and epistemic uncertainty

The preceding diagram delineates two distinct point regions. The left region exhibits pronounced randomness, while the right region showcases a structured data pattern, evident from a straight line that can be drawn through its clustered points. The left cluster demonstrates high aleatoric uncertainty, in contrast to the low aleatoric uncertainty in the right cluster, attributed to its data regularities. Additionally, three areas manifesting high epistemic uncertainty correspond to the voids in data, indicating gaps in our understanding or knowledge of the system.

Let's move on to cover different ways to quantify uncertainty.

Different ways to quantify uncertainty

There are several different approaches to quantify uncertainty, each with its own strengths and limitations. Here are a few examples:

- **Statistical methods**: Statistical methods are widely used for UQ and involve using probability distributions to model the uncertainty in data and predictions. These methods are widely used in fields such as finance, engineering, and physics and involve tools such as confidence intervals, regression analysis, Monte Carlo simulations and hypothesis testing.

- **Bayesian methods**: Bayesian methods involve using prior knowledge and data to update our beliefs about the uncertainty in predictions. These methods are widely used in machine learning, natural language processing, and image processing. Bayesian tools include Bayesian inference – statistical methods to update beliefs about the uncertainty of predictions based on new data – and Bayesian networks – graphical models that represent probability relationships between the variables that can be used to model various systems and calculate the probabilities of different outcomes. Bayesian tools also include **Markov Chain Monte Carlo** (**MCMC**) – a computational method used to sample from complex probability distributions – as well as Bayesian optimization – methods used to optimize a function that is expensive to evaluate.

- **Fuzzy logic methods**: Fuzzy logic involves using sets and membership functions to represent uncertainty in a system. This approach is widely used in control systems, robotics, and artificial intelligence. Fuzzy logic includes fuzzy sets – that is, sets that allow partial membership. Rather than a binary classification of an element as either a member or non-member of a set, fuzzy sets allow elements to have degrees of membership. This allows uncertainty to be represented in a more nuanced way.

We will now talk about quantifying uncertainty using conformal prediction.

Quantifying uncertainty using conformal prediction

Quantifying the uncertainty of machine learning predictions is becoming increasingly important as machine learning is used more widely in critical applications such as healthcare, finance, and self-driving cars. In these applications, the consequences of incorrect predictions can be severe, making it essential to understand the uncertainty associated with each prediction.

For example, in healthcare, machine learning models are used to make predictions about patient outcomes, such as the likelihood of a disease or the effectiveness of a treatment. These predictions can have a significant impact on patient care and treatment decisions. However, if the model is unable to produce an estimate of its own confidence, it may not be useful and could potentially be risky to rely upon.

In contrast, if the model can provide a measure of its own uncertainty, clinicians can use this information to make more informed decisions about patient care and treatment.

Consider a situation where a doctor has obtained a patient's MRI scan and has to deduce whether the patient has cancer. In this application, high accuracy is not enough, as the doctor's diagnosis has to rule out (or not) a patient having such a critical, life-changing disease as cancer.

Similarly, in finance, machine learning models are used to make predictions about market trends, stock prices, and risk assessments. These predictions can have a significant impact on investment decisions and portfolio management. However, if the model is unable to produce an estimate of its own confidence, it may not be useful and could potentially lead to poor investment decisions. On the other hand, if the model can provide a measure of its own uncertainty, investors can use this information to make more informed decisions and reduce the risks associated with uncertainty.

Quantifying uncertainty is a prerequisite for explainability and trust in machine learning models. If a model cannot provide an estimate of its own confidence, it may be difficult to explain why it made a certain prediction or to gain the trust of stakeholders.

In contrast, if the model can provide a measure of its own uncertainty, stakeholders can better understand how the model arrived at its predictions and can have more confidence in the model's performance.

Conformal prediction is a relatively new framework for quantifying uncertainty in predictions that offers several advantages over traditional statistical, Bayesian, and fuzzy logic methods.

Here are some advantages of using conformal prediction for UQ:

- **Probabilistic predictions**: Conformal prediction provides probabilistic predictions that include measures of confidence, accuracy, and reliability for each prediction. This allows users to make informed choices and to estimate the range of possible outcomes.

- **Coverage guarantees**: Unlike other methods, prediction regions generated by conformal prediction models (prediction sets for classification tasks/predictive intervals for regression tasks) come with rigorous statistical validity guarantees. This valuable information can be used to assess the reliability of the model, mitigate the risks, and identify areas where further research or data collection is needed.

- **Non-parametric**: Conformal prediction does not require assumptions about the underlying probability distributions, making it applicable to a wide range of problems and data types.

- **Distribution agnostic**: Conformal prediction models work for any data distribution as long as exchangeability assumptions can be maintained. However, conformal prediction models have recently been extended to contexts where the exchangeability assumption is no longer met, including successful applications in time series forecasting.

- **No restrictions on dataset size**: Unlike statistical and Bayesian models, conformal prediction models are not concerned with dataset size – the validity of predictions is maintained regardless of the size of the dataset. However, prediction intervals are usually narrower where there is a larger amount of data. This is due to the general pattern of machine learning models being able to learn more effectively from more data.

- **Robustness**: Conformal prediction is robust to outliers and noisy data, making it particularly useful in settings where the data may be imperfect or incomplete.

- **Wide application**: Conformal prediction has been successfully applied to a wide variety of problem classes, including classification, regression, time series and forecasting, computer vision, NLP, reinforcement learning, and much more.

- **Low computation cost**: Conformal prediction models, while offering very rigorous prediction regions for UQ, maintain computational efficiency and don't heavily burden system resources, making them ideal for real-time applications and large datasets.

In contrast to alternative approaches, conformal prediction offers robust and non-parametric probabilistic predictions with assured validity. This becomes especially beneficial in scenarios where analytical modeling of uncertainty is challenging and probabilistic predictions are needed for decision making.

Conformal prediction quantifies uncertainty by offering a probability measure that indicates the likelihood of a prediction being accurate. This uncertainty measure is rooted in the notion of validity, which denotes the expected ratio of correct predictions.

In classification tasks, a machine learning model typically produces class scores and assigns the label with the highest score. However, this can pose issues when the prediction certainty is low.

Consider the following images from a *Gentle Introduction to Conformal Prediction and Distribution-Free Uncertainty* (https://arxiv.org/abs/2107.07511) by Anastasios N. Angelopoulos and Stephen Bates. A trained deep learning model will output class scores for new images just like the three in the following figure.

We can see that these three examples do not produce the same results.

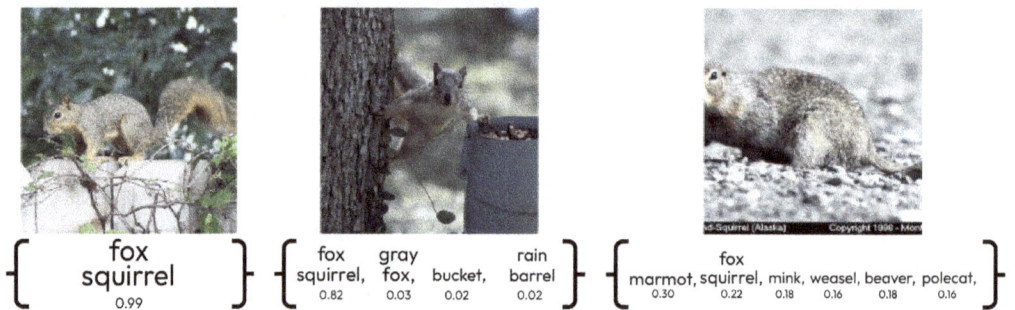

Figure 2.2 – Uncertainty in a classification problem

> **Note**
> The preceding figure is sourced from *Angelopoulos and Bates, 2023* (used with permission by the authors).

The left-hand picture is easy to predict – the model would have seen a lot of examples of squirrels and is rather certain about its prediction. As a result, the prediction set contains just one potential label – fox squirrel.

The middle picture has a medium level of uncertainty associated with it – while there are additional objects in the background, the model is still quite sure that it is most likely a fox squirrel. However, to hedge its predictions, the model now outputs additional labels such as *gray fox, bucket, and rain barrel*.

The picture on the right is the hardest to classify as there is a lot of uncertainty in the dataset; the head of the animal is partially occluded and the background is unusual. To account for this prediction uncertainty, the model has not only reduced the probability of the `fox squirrel` class but also introduced the possibility that the animal is in fact a marmot. In addition, the model has extended the prediction set by leaving open the possibility that this animal can in fact also be *a mink, weasel, beaver, or even a polecat*.

Note how the conformal prediction model extended the point prediction of the deep neural network as, instead of generating a point prediction, it produced a prediction set given the specified 95% confidence level. The prediction set produced by the conformal prediction model provides much more information to improve downstream decision making.

The conformal prediction framework is based on the idea of constructing a set of predictions that includes the true value with a certain degree of confidence. This set is known as a conformal prediction set and is constructed based on a set of training data.

To construct a conformal prediction set, the framework uses a non-conformity function that measures how well a prediction fits the training data.

The non-conformity measure is used to rank the predictions in order of how well they fit the training data. The most similar predictions are included in the conformal prediction set, along with a set of additional predictions that have a certain degree of confidence based on the non-conformity measure. By following this approach, the conformal prediction model can provide a measure of its own uncertainty. This allows both the modelers and the stakeholders to better understand how the model arrived at its predictions and can engender more confidence in the model's performance.

Summary

In this chapter, we have provided an overview of conformal prediction and explained why conformal prediction is a valuable tool for quantifying the uncertainty of predictions, especially in critical settings such as healthcare, self-driving cars, and finance. We also discussed the concept of UQ and how the conformal prediction framework has successfully addressed the challenge of quantifying uncertainty.

In the next chapter, we will dive deeper into the fundamentals of conformal prediction and apply it to binary classification problems. We will illustrate how you can apply conformal prediction to your own binary classification problems by computing non conformity scores and p-values and then using the p-values to decide which class labels should be included in your prediction sets.

Part 2:
Conformal Prediction
Framework

This part will explain the fundamentals of conformal prediction. You will learn about the types of conformal prediction models and the critical concepts of conformal prediction, including validity, efficiency, and non-conformity measures.

This section has the following chapters:

3

Fundamentals of Conformal Prediction

This chapter will dive into **conformal prediction**, a powerful and versatile probabilistic prediction framework. Conformal prediction allows for effective quantification of uncertainty in machine learning applications. By learning and utilizing conformal prediction techniques, you will be able to make more informed decisions and manage risks associated with data-driven solutions more effectively.

This chapter will cover the mathematical underpinnings of conformal prediction. You will learn how to accurately measure the uncertainty that comes with your predictions. You will also become familiar with nonconformity measures, grasp the idea of prediction sets, and be able to evaluate your model's performance in a thorough and meaningful manner. The abilities you will acquire through this chapter will be highly valuable in various academic and industrial fields where comprehending the uncertainty associated with predictions is essential.

In this chapter, we're going to cover the following main topics:

- Fundamentals of conformal prediction
- Basic components of a conformal predictor

By mastering the concepts and techniques presented in this chapter, you will be well equipped to harness the power of conformal prediction and effectively apply it to your industrial applications.

Fundamentals of conformal prediction

In this section, we will cover the fundamentals of conformal prediction. There are two variants of conformal prediction – **inductive conformal prediction** (ICP) and **transductive conformal prediction** (TCP). We will discuss the benefits of the conformal prediction framework and learn about the basic components of conformal predictors and the different types of nonconformity measures. We will also learn how to use nonconformity measures to create probabilistic prediction sets in classification tasks.

Definition and principles

Conformal prediction is a machine learning framework that quantifies uncertainty to produce probabilistic predictions. These predictions can be prediction sets for classification tasks or prediction intervals for regression tasks. Conformal prediction has significant advantages in equipping statistical, machine learning, and deep learning models with valuable additional features that instill confidence in their predictions.

Moreover, it is the only uncertainty quantification framework that offers strong mathematical assurances that the error rate will never surpass the significance level determined by the user. Simply put, conformal prediction models always generate valid and unbiased prediction sets and prediction intervals, which is a crucial aspect of making informed decisions.

For comprehensive resources, visit *Awesome Conformal Prediction* (`https://github.com/valeman/awesome-conformal-prediction`). It is the most extensive, professionally curated resource on conformal prediction. Over time, conformal prediction has developed into an extensive framework suitable for use with any underlying point prediction model, regardless of the size of the dataset, the underlying point prediction model, or the data distribution.

Today, the principles of conformal prediction can be expressed as follows:

- **Validity**: The objective of conformal prediction is to create **prediction regions** (such as **prediction sets** for classification tasks or **prediction intervals** for regression tasks) that encompass the actual target value with a confidence level specified by the user. The goal is to attain a coverage probability at least as large as the user-defined confidence level. For example, if the user selects a 95% confidence level, the prediction regions (sets or intervals) should contain the correct target values at least 95% of the time.

- **Efficiency**: Conformal prediction aims to generate prediction intervals or regions that are as small as possible while preserving the desired confidence level. This approach ensures the predictions are valid and precise while conveying useful information.

- **Adaptivity**: Conformal prediction aims to generate prediction sets that are adaptive to individual examples. For examples that are hard to predict, prediction sets are expected to be wider to account for uncertainty in predictions.

- **Distribution-free**: Conformal prediction is a versatile and robust framework that can be utilized with various data types and machine learning tasks since it does not depend on specific assumptions about the underlying data distribution. The only assumption made by conformal prediction is data exchangeability, which is a less restrictive requirement than that of **independent, identically distributed** (**IID**) data. Nevertheless, conformal prediction has succeeded in numerous applications beyond exchangeability, including time series and forecasting applications.

- **Online adaptivity**: Conformal prediction can operate in both online and offline scenarios. In online settings, a conformal predictor can adjust to new incoming data points and modify its predictions accordingly without retraining the model.

- **Compatibility**: Conformal prediction is a flexible framework that can be seamlessly integrated with various statistical and machine learning techniques, including decision trees, neural networks, support vector machines, boosted trees (XGBoost/LightGBM/CatBoost), bagging trees (random forest), and deep learning models. This is achieved by defining an appropriate nonconformity measure that can be applied to any existing machine learning algorithm.

- **Non-intrusive**: Conformal prediction does not necessitate statistical, machine learning, or deep learning point prediction model changes. This is particularly important for models that have been deployed into production. Conformal prediction can be added as an uncertainty quantification layer on top of any deployed model without any modification or knowledge of how the prediction model functions.

- **Interpretability**: Conformal prediction produces prediction sets and intervals that are easy to understand and provide a clear way to measure uncertainty. This makes it a valuable tool for industries such as finance, healthcare, and autonomous vehicles, where understanding prediction uncertainty is crucial.

In the next section, we will look at the basic components of a conformal predictor and learn about nonconformity measures.

Basic components of a conformal predictor

We will now look at the basic components of a conformal predictor:

- **Nonconformity measure**: The nonconformity measure is a function that evaluates how much a new data point differs from the existing data points. It compares the new observation to either the entire dataset (in the full transductive version of conformal prediction) or the calibration set (in the most popular variant – **ICP**. The selection of the nonconformity measure is based on a particular machine learning task, such as classification, regression, or time series forecasting, as well as the underlying model. This chapter will examine several nonconformity measures suitable for classification and regression tasks.

- **Calibration set**: The calibration set is a portion of the dataset used to calculate nonconformity scores for the known data points. These scores are a reference for establishing prediction intervals or regions for new test data points. The calibration set should be a representative sample of the entire data distribution and is typically randomly selected. The calibration set should contain a sufficient number of data points (at least 500). If the dataset is small and insufficient to reserve enough data for the calibration set, the user should consider other variants of conformal prediction – including **TCP** (see, for example, *Mastering Classical Transductive Conformal Prediction in Action* – `https://medium.com/@valeman/how-to-use-full-transductive-conformal-prediction-7ed54dc6b72b`).

- **Test set**: The test set contains new data points for generating predictions. For every data point in the test set, the conformal prediction model calculates a nonconformity score using the nonconformity measure and compares it to the scores from the calibration set. Using this comparison, the conformal predictor generates a prediction region that includes the target value with a user-defined confidence level.

All these components work in tandem to create a conformal prediction framework that facilitates valid and efficient uncertainty quantification in a wide range of machine learning tasks.

In the previous chapter, we examined nonconformity measures, which are the fundamental element of any conformal prediction model. To recap, the primary role of the nonconformity measure is to enable the quantification of uncertainty for new data points by evaluating the extent to which they differ from previously observed data.

In the conformal prediction framework, any model provides valid prediction sets regardless of the chosen nonconformity measure. However, selecting the proper nonconformity measure is essential for creating more precise, informative, and adaptive prediction regions.

In conformal prediction, the size of these regions determines the effectiveness of predictive systems. Smaller regions are considered more efficient and informative.

The effectiveness of a conformal prediction model can be influenced by the nonconformity measure that is chosen by the user. However, the best nonconformity measure for machine learning is dependent on the context.

For further understanding, you can explore the *no free lunch* theorem, which discusses the limitations of universal optimization and learning algorithms. You can find more information on this theorem at (`https://en.wikipedia.org/wiki/No_free_lunch_theorem`).

Despite the theorem's implications, research papers have provided valuable insights into selecting effective nonconformity measures. These papers have examined diverse datasets and offer guidance on choosing nonconformity measures that have demonstrated effectiveness across various scenarios. By leveraging these research findings, you can make informed decisions when selecting nonconformity measures for your specific machine learning tasks.

In the research paper titled *Model-Agnostic Nonconformity Functions for Conformal Classification* (`https://ieeexplore.ieee.org/abstract/document/7966105`), the authors examined the efficiency of three model-agnostic nonconformity measures for classification problems. In the experiments on 21 multi-class datasets using neural networks and neural network ensembles as classifiers, the authors discovered that the choice of the nonconformity measure substantially influenced the efficiency of prediction sets. These findings highlight the importance of selecting an appropriate nonconformity measure to enhance the efficiency of conformal prediction in classification problems.

The researchers concluded that choosing the optimal nonconformity measure depends on the efficiency metric most suitable for the use case. When evaluating the efficiency in terms of the proportion of single-label predictions (singletons), the margin-based nonconformity measure emerged as the preferred option. On the other hand, when assessing the average width of prediction sets, the hinge loss measure resulted in the narrowest prediction sets, indicating its effectiveness in producing more precise and focused predictions.

In conformal prediction, the nonconformity measures are derived from the predictions generated by the underlying point prediction model. Instances more prone to misclassification or greater inherent uncertainty are assigned higher nonconformity scores. The efficiency of a conformal prediction model depends on both the accuracy of the underlying model and the quality of the chosen nonconformity measure. Selecting an appropriate nonconformity measure becomes particularly crucial when dealing with challenging datasets where the underlying point prediction model needs additional support to classify objects accurately.

Types of nonconformity measures

There are two types of nonconformity measures – **model-dependent** and **model-independent** ones.

Model-dependent nonconformity measures are specific to a particular type of underlying model used in conformal prediction. These measures rely on the internal workings or characteristics of the model to compute nonconformity scores. Unlike model-agnostic nonconformity measures, which can be applied to any type of point prediction model, model-dependent measures are tailored to the specific model used.

Model-dependent nonconformity measures take advantage of the unique features or properties of the underlying model to assess the deviation or uncertainty of a new data point from the training (in classical TCP) or calibration (in ICP) data. These measures can be customized based on the model's output, such as the probability estimates or decision boundaries. They can also leverage model-specific attributes, such as the learned weights or parameters, to determine the nonconformity scores.

Examples of model-dependent nonconformity measures include the distance to support vectors in support vector machines, the residual error in linear regression models, or the discrepancy between predicted and actual class probabilities in probabilistic classifiers.

Model-dependent nonconformity measures offer the advantage of potentially capturing model-specific information and characteristics, leading to tailored and potentially more accurate uncertainty quantification. However, they are limited to the specific model they are designed for and may not generalize well to other models. For this reason, we will only cover model-independent nonconformity measures.

In *Model-Agnostic Nonconformity Functions for Conformal Classification*, the authors investigated three popular loss functions – hinge loss, margin, and Brier score – as popular choices of model-independent nonconformity measures for predictive classification. Since these functions work with any classification model producing class estimates, they can be utilized with any classifier that generates class scores, making them model-agnostic.

It is important to note that, in a conformal predictor, only the ordering of nonconformity scores matters. In two-class problems, all three loss functions – hinge loss, margin, and Brier score – will arrange the instances in the same order, resulting in the same efficiency.

To compare these efficiency measures, the authors focused on multi-class problems.

Let's look at the three nonconformity measures in more detail.

Hinge loss

The hinge loss (also sometimes called **LAC loss** or **inverse probability**) can be described in the context of classification problems where we obtain class probabilities as outputs from a model.

For a particular instance, let's say the true label is y and the predicted probability of the model for that label is $P(y)$. Then, the hinge loss for this instance is calculated as:

$$\text{Hinge loss} = 1 - P(y)$$

Explanation

If the model is very confident and assigns a probability of 1 to the true label y, then the hinge loss is 0. This indicates a perfect prediction.

If the model assigns a probability of 0 to the true label y, then the hinge loss is 1. This indicates a completely incorrect prediction.

For probabilities between 0 and 1, the hinge loss will range between 0 and 1, with higher values indicating lower confidence in the correct label and vice versa.

In essence, the hinge loss gives a measure of how "off" the prediction is from the true label. Lower hinge loss values are better, indicating that the predicted probability for the true label is closer to 1. Conversely, higher hinge loss values indicate greater disagreement between the predicted probability and the true label.

To illustrate how to compute the hinge loss nonconformity score, consider a classifier that produces three class scores: *class_0 = 0.5*, *class_1 = 0.3*, and *class_2 = 0.2*, and the actual label $y = 1$.

To compute the nonconformity score, take the probability score of the true class (in this case, 1) and subtract it from 1. Thus, this example's inverse probability (hinge) nonconformity score is 0.7.

Margin nonconformity measure

The margin nonconformity measure is defined as the difference between the predicted probability of the most likely incorrect class label and the predicted probability of the true label:

$$\Delta\left[h(x_i), y_i\right] = \max_{y \neq y_i} \hat{P}_h(y \mid x_i) - \hat{P}_h(y_i \mid x_i)$$

Explanation

The measure captures the difference in probabilities between the highest probability given to any incorrect label and the probability of the true label y_i for the instance x_i. For a particular instance x_i with true label y_i, we first identify the probability of the most likely incorrect class and then subtract from this value the probability of the true label to get the margin.

If the margin is close to zero or negative, it means that the model is confident in its prediction for the true class label and there isn't another class with a closely competing probability. This indicates a conforming example.

If the margin is positive and large, it indicates that there's another class (an incorrect one) for which the model assigns a higher probability than the true class. This is a nonconforming example, indicating that the model is more confident in an incorrect class than in the true one.

The larger the margin, the more nonconforming the example is, as it suggests greater disagreement between the predicted probabilities for the true class and the most likely incorrect class.

In essence, the margin-based nonconformity measure gives an indication of how "risky" a prediction is. If the measure is high, it indicates potential problems with the model's prediction for that instance, signaling that the prediction might be unreliable. If the measure is close to zero or negative, it means the model is more confident in its prediction of the true class.

To illustrate how to compute the margin nonconformity score, consider a classifier that produces three class scores: *class_0 = 0.5, class_1 = 0.3, and class_2 = 0.2*, and the actual label *y = 1*.

To calculate the margin nonconformity score, one would take the probability of the most likely but incorrect class (in this case, 0) and subtract it from the probability of the true class (1) to get a margin nonconformity score of 0.2.

The Brier score

The Brier score measures the accuracy of probability-based predictions in classification tasks. It calculates the squared difference between the predicted probabilities and the actual binary results. The score's values can range from 0 (perfect accuracy) to 1 (complete inaccuracy).

The Brier score is an example of a **proper scoring rule** (another example of a proper scoring rule in classification problems is log loss).

A proper scoring rule is a metric used to evaluate the accuracy of probabilistic predictions. Specifically, it's a rule that assigns a numerical score to each prediction in such a way that the most accurate (or calibrated) probabilistic forecast will, on average, receive a better (typically lower) score than any other biased or less accurate forecast.

The idea behind a proper scoring rule is to encourage honest reporting of probabilities. If a scoring rule is "proper," then a forecaster has the best expected score when they report their true beliefs or true estimated probabilities, rather than exaggerating or downplaying their forecasts.

In essence, a proper scoring rule ensures that forecasters are best rewarded, in terms of the score, when they provide their genuine assessments of the probabilities of events.

As a proper scoring rule, the Brier score promotes well-calibrated probability estimates. It uniquely captures both the calibration, which refers to the alignment between predicted probabilities and actual outcomes, and the discrimination, or the model's ability to distinguish between the classes.

Glenn W. Brier, who worked in weather forecasting, invented the Brier score in the 1950s and described it in his paper *Verification of forecasts expressed in terms of probability* (https://journals. ametsoc.org/view/journals/mwre/78/1/1520-0493_1950_078_0001_vofeit_2_0_ co_2.xml). Along with log loss, the Brier score is widely used today to evaluate the performance of probabilistic classifiers and understand the quality of predicted probabilities.

To illustrate how to compute the Brier score using our example, consider the same classifier, which produces three class scores: *class_0 = 0.5, class_1 = 0.3, and class_2 = 0.2*, and the actual label *y = 1*. Here's a step-by-step guide to how to compute the Brier score for the given example:

1. *Encode the actual class*:

    For a multi-class problem, you'll want to use one-hot encoding for the actual labels:

    - *class_0*: 0

    - *class_1*: 1

    - *class_2*: 0

2. *Compute the squared differences for each class*:

    Calculate the squared difference between predicted probabilities and the actual outcomes:

    - *For class 0: $(0-0.5)^2 = 0.25$*

    - *For class 1: $(1-0.3)^2 = 0.49$*

    - *For class 2: $(0-0.2)^2 = 0.04$*

3. *Average the squared differences*:

$$\text{Brier score} = \frac{0.25 + 0.49 + 0.04}{3} = 0.26$$

Thus, the Brier score for this example is 0.26. A lower Brier score indicates better performance, with 0 being the best possible score.

Effectiveness of model-agnostic nonconformity measures

In the paper *Model-Agnostic Nonconformity Functions for Conformal Classification*, the authors assessed the effectiveness of three nonconformity measures using two criteria: **one-class classification (OneC)** and *AvgC*.

OneC refers to the proportion of all predictions that consist of singleton sets containing only one label. These sets are desired because they provide the most informative predictions.

AvgC, on the other hand, refers to the average number of class labels in the prediction set. A lower **AvgC** value indicates that the model is better at producing more specific and informative predictions by eliminating class labels that do not fit well.

For example, consider a binary classification problem with the following predictions for five instances:

- *Prediction Set 1: {0}*
- *Prediction Set 2: {1}*
- *Prediction Set 3: {0, 1}*
- *Prediction Set 4: {1}*
- *Prediction Set 5: {0}*

In this case, there are five predictions, four of which are singleton sets (prediction sets 1, 2, 4, and 5). To calculate **OneC**, we compute the proportion of singleton sets among all predictions:

OneC = (Number of Singleton Sets) / (Total Number of Prediction Sets) = (4) / (5) = 0.8

A higher **OneC** value indicates that the conformal prediction model produces specific and informative predictions more efficiently. In this example, 80% of the prediction sets are singletons, reflecting a relatively efficient classifier.

To calculate **AvgC**, which represents the average number of class labels in the prediction set, we first compute the sum of the class labels in each prediction set:

- **Prediction Set 1**: 1 label
- **Prediction Set 2**: 1 label
- **Prediction Set 3**: 2 labels
- **Prediction Set 4**: 1 label
- **Prediction Set 5**: 1 label

The sum of the class labels is $1 + 1 + 2 + 1 + 1 = 6$.

Next, we divide this sum by the total number of prediction sets:

AvgC = (Sum of Class Labels) / (Total Number of Prediction Sets) = (6) / (5) = 1.2

In this example, **AvgC** is 1.2, indicating that, on average, each prediction set contains 1.2 class labels. A lower **AvgC** value signifies that the model is better at producing more specific and informative predictions. In this case, the **AvgC** value of 1.2 reflects a relatively efficient classifier, as it is close to the minimum possible value of 1, which would occur if all prediction sets were singleton sets.

Researchers have determined that the most effective approach is to use a margin-based nonconformity function to achieve a high rate of singleton predictions (**OneC**).

On the other hand, a nonconformity measure utilizing the hinge (inverse probability) nonconformity measure yielded the smallest label sets on average, as measured by **AvgC**.

What is the intuition behind such results?

To achieve a high **OneC** score, the predictions should predominantly be singleton sets that contain only one label. This means that high nonconformity scores must be assigned to all other labels. The margin-based nonconformity measure fosters this outcome, especially when the underlying model attributes a high probability to a single label. In this scenario, the probability associated with that single label is added to the nonconformity scores for all other labels, effectively promoting the selection of a singleton set and thereby improving the **OneC** performance.

However, the hinge loss function, which assesses each label on an individual basis, might only exclude some labels, leaving the high-probability ones intact in certain cases. This situation arises because all other labels must have inherently low probabilities, and the hinge loss function does not take into account how the remaining probability mass is distributed specifically to the high-probability label. Consequently, the hinge loss function's inability to consider this distribution may lead to it only eliminating some labels and not necessarily focusing on the high-probability ones. This difference in how the two nonconformity measures assign scores leads to margin-based nonconformity measures being better suited for achieving a high proportion of singleton predictions.

On the other hand, to obtain smaller label sets on average as measured by **AvgC**, a nonconformity measure based on the hinge (inverse probability) nonconformity measure is more effective. This is because hinge loss considers only the probability of the true class label. In contrast, the margin-based nonconformity measure considers both the probability of the true class label and the most likely incorrect class label. This leads to the margin-based nonconformity measure producing broader sets on average, whereas hinge loss is more likely to eliminate incorrect labels and produce smaller, more informative sets.

Compared to classification, regression problems are relatively straightforward regarding the selection of nonconformity measures:

- **Absolute error**: The absolute error nonconformity measure is the absolute difference between the predicted value and the true target value for a given data point. This measure can be used with any regression model:

$$Nonconformity\ (x) = |y\_pred - y\_true|$$

- **Normalized error**: The normalized error nonconformity measure is the absolute error divided by an estimate of the prediction error's scale, such as the **mean absolute error** (**MAE**) or the standard deviation of the residuals. This measure can be used with any regression model and helps account for heteroscedasticity in the data:

$$Nonconformity\ (x) = |y\_pred - y\_true|\ /\ scale$$

Let's now consider the pros and cons of both nonconformity measures in the regression problem.

Absolute error

The pros are given as follows:

- **Simplicity**: It is straightforward to compute and understand, making it a go-to choice for many practitioners

- **Uniform interpretation**: Given that it is not scaled, the interpretation remains consistent across different datasets

The cons are given as follows:

- **Scale sensitivity**: The absolute error can be sensitive to the scale of the target variable. For datasets with large target values, the absolute error might be large, even if the predictions are relatively accurate.

- **No consideration for data distribution**: It does not consider the variability or distribution of errors in the dataset, which might lead to overly optimistic or pessimistic conformal predictions.

Normalized error

The pros are given as follows:

- **Scale invariance**: By normalizing the error with respect to the error's scale (e.g., MAE or standard deviation of residuals), it becomes less sensitive to the scale of the target variable, allowing for more consistent performance across datasets with varying scales.

- **Accounts for heteroscedasticity**: This measure can be particularly useful for data that exhibits heteroscedasticity (i.e., where the variability of errors changes across the data). By normalizing with a measure of spread, it can give a more accurate representation of the prediction's relative accuracy.

- **More adaptive**: The normalization factor can adapt to the local properties of the data, providing more meaningful error measurements.

The cons are given as follows:

- **Complexity**: It introduces an additional layer of complexity as one needs to determine the best way to normalize the errors. The choice of normalization (e.g., using MAE versus standard deviation of residuals) can influence the results.

- **Risk of misleading results**: If the normalization factor is not well chosen or if it is computed from a small sample, it might lead to misleading conformal predictions.

- **Requires more data**: Estimating the scale of prediction error typically requires a sufficiently large sample size to be reliable.

In the context of conformal prediction for regression, the choice between these measures often depends on the properties of the data and the specific needs of the application. If heteroscedasticity is a concern, the normalized error might be more appropriate. Otherwise, the simplicity of the absolute error might be preferred.

The second component of a conformal predictor is the **calibration set**, which is used to compute nonconformity scores for the known data points. The calibration set is a feature of the most popular variant, ICP, while TCP does not require a calibration set. In contrast to ICP, TCP utilizes all available data, making it efficient regarding data usage. However, TCP is computationally inefficient, requiring retraining the underlying point prediction model for each new test object.

We will focus on ICP, the most popular and widely used variant in current research and open source libraries, to better understand conformal prediction. ICP was introduced in a 2002 paper *Inductive Confidence Machines for Regression* (`https://link.springer.com/chapter/10.1007/3-540-36755-1_29`).

ICP has a significant advantage in terms of computational efficiency, as it is almost as fast as the underlying point prediction model. This is because ICP generates a single model, based on the training data, which can then be used to produce predictions for all test instances. Any predictive model can be combined with ICP to convert it into a conformal predictor.

When developing an ICP, remember that the training set is exclusively for training the base prediction model. Do not use it to construct the conformal predictor. Likewise, the calibration set should be reserved solely for the conformal predictor and not for training the base model.

The main objective of conformal prediction is to provide valid prediction sets for new, unseen examples. The ICP approach enables the model to learn about uncertainty by comparing predictions made by the underlying point prediction model with the actual labels.

The ICP is constructed as follows:

1. Divide your training data into two disjoint subsets I_T and I_C where T and C denote the proper training and calibration sets.

2. Train your point prediction model, H, using data exclusively from the appropriate proper training set, I_T. As discussed in the previous chapter, H can be any point prediction model, including statistical, machine learning, deep learning, or even any model based on expert opinions, business rules, or heuristics.

3. Use an appropriate nonconformity measure for your classification or regression task to calculate nonconformity scores $\alpha 1, \alpha 2, ..., \alpha n$, where n represents the total number of data points in the calibration dataset.

4. Tentatively assign a label y as the potential label for the new test point x, and compute the nonconformity score α for (x,y).

5. Calculate the p-value as follows:

$$p = \frac{|\{z_i : \alpha_i \geq : \alpha_\tau\}| + 1}{n + 1}.$$

Let's briefly discuss this important step of calculating the p-value for each test data point. In previous steps, we computed the nonconformity scores for all points in the calibration set. We then compute the nonconformity score of a new test point to determine how the test point's nonconformity score compares to those in the calibration set. The p-value provides a measure of how different the test object is from the calibration data, based on its nonconformity score. It is calculated by first calculating a nonconformity score for the test object using the model, then calculating nonconformity scores for all objects in the calibration set using the same model. The number of calibration objects that have a nonconformity score greater than or equal to the test object's score is then counted, and 1 is added to this count. This count plus 1 is then divided by the total number of calibration objects plus 1 to determine the p-value, which indicates the proportion of calibration objects that are at least as extreme as the test object. The resulting p-value provides a quantitative metric of how nonconforming the test object is to the pattern in the calibration data, with a low p-value meaning the test object is highly unusual compared to the calibration data.

6. Conformal prediction's core idea is to assign a tentative label y to the test point (a class label in classification problems or a real y value in regression problems) and evaluate how well the test object, including its features and assigned label, fits in with the observed objects from the calibration set. To measure this fit, we compute the p-value by comparing the test object's "strangeness" using the test object's nonconformity score to that of the calibration set objects.

    It is essential to note that when calculating the p-value, the numerator and denominator (n+1) include the test object in the same bag, together with the calibration set data. Due to the exchangeability assumption, data can be shuffled, making all data points equivalent in terms of their order.

    Once we have the p-value for the test point with a tentatively assigned label, we compare it to the significance level. If the p-value is lower than the significance level, it indicates that very few, if any, objects in the calibration set are as strange as our test point. This suggests that the proposed label y does not fit, and we exclude it from the prediction set.

    On the other hand, if the p-value is equal to or higher than the significance level, assigning the potential label y would not make our test object particularly strange, given the observed data. In this sense, we use p-values to test the statistical hypothesis of whether each potential y value fits previously observed data given the exchangeability assumption.

7. We repeat the process mentioned above for each possible value of y. As a result, we obtain a prediction set that includes the true label with a probability of 1 - ε, where ε is our chosen significance level (for example, 5%).

We will now discuss the concepts of confidence and credibility, which help assess the quality of predictions:

- **Confidence level**: The confidence level, denoted by $(1 - \varepsilon)$, represents the probability with which the true label (or value) falls within the prediction set. A higher confidence level indicates that the predictions are more likely to be accurate. The confidence level is usually chosen in advance, and common values are 0.95 (95%) or 0.99 (99%). With a 95% confidence level, for example, one can expect that the true label will be included in the prediction set 95% of the time.

- **Credibility level**: The credibility level, denoted by p, measures the likelihood of each element within the prediction set being the true label. We have discussed calculating p-values for each potential value of label 'y'. In classification tasks, credibility levels can be interpreted as a normalized measure of confidence in each class label. Credibility levels help determine the most likely value(s) within the prediction interval for regression tasks. The credibility level can be used as a threshold to filter out less probable predictions, leading to a more precise, albeit smaller, prediction set.

Let's consider a binary classification problem with two possible class labels: *A* and *B*. We'll use a conformal prediction framework to calculate confidence and credibility levels for a test data point.

Assume we have already trained a point prediction model and calculated the nonconformity scores for the calibration dataset and the test point.

Nonconformity scores for the calibration dataset are as follows:

- *Point 1 (Label A): 0.4*
- *Point 2 (Label A): 0.3*
- *Point 3 (Label B): 0.2*
- *Point 4 (Label B): 0.5*

Nonconformity scores are calculated for a test point using the model's predicted labels and probabilities. For example, if the model produces probability estimates of 0.75 for class A and 0.65 for class B, nonconformity scores would be computed as follows:

- For class A, the model assigns a probability of 0.75. The nonconformity score is calculated by subtracting this probability from 1, giving $1-0.75 = 0.25$.
- For class B, the model assigns a probability of 0.65. The nonconformity score is $1-0.65 = 0.35$.

In general, the nonconformity score is 1 minus the model's probability for the tentative label. This measures how much the prediction deviates from full confidence (a probability of 1). Higher nonconformity scores mean the model's label assignments are less certain or conforming. Hinge loss is commonly used when the model outputs probability estimates for each class. Subtracting the probability from 1 provides an intuitive nonconformity measure.

Now, we'll calculate p-values for each tentative label:

1. *Label A*: Count the number of calibration points with nonconformity scores greater than or equal to 0.25: 3 (Points 1, 2, and 4)

$$p-value = (3 + 1) / (4 + 1) = 4 / 5 = 0.8$$

2. *Label B*: Count the number of calibration points with nonconformity scores greater than or equal to 0.35: 2 (Points 1 and 4)

$$p-value = (2 + 1) / (4 + 1) = 3 / 5 = 0.6$$

These p-values are the credibility levels for each label. Thus, the credibility of label A is 0.8, and the credibility of label B is 0.6.

Our desired confidence level is 95% ($1-\varepsilon = 0.95$, $\varepsilon = 0.05$). Since both credibility levels (p-values) are greater than the chosen significance level $\varepsilon = 0.05$, we include both labels A and B in the prediction set.

In this example, the confidence level is 95%, indicating that the true label is included in the prediction set with a 95% probability. The credibility levels are 0.8 for label A and 0.6 for label B, suggesting that label A is more likely to be the correct label than label B.

However, both labels are included in the prediction set because their credibility levels are higher than the significance level.

Online and offline conformal prediction are two variants of conformal prediction that differ in how they process and incorporate new data points:

- **Offline conformal prediction**: In offline conformal prediction, a model is trained on a fixed dataset, and the nonconformity scores are calculated for a separate calibration dataset. The model does not update or change as new data points become available. This method is suitable when the dataset is static or when you have a large amount of data available for training and calibration. The disadvantage of offline conformal prediction is that it doesn't adapt to new data or changes in data distribution over time.

- **Online conformal prediction**: In online conformal prediction, the model continuously updates as new data points become available. It incorporates new information by updating the nonconformity scores and adjusting the predictions accordingly. Online conformal prediction is particularly useful when working with streaming data or when the underlying data distribution changes over time. This method lets the model stay up to date with the most recent data, providing more accurate predictions in dynamic environments. However, online conformal prediction can be computationally more demanding due to the need for constant updates.

Conditional and unconditional coverage are two criteria used to evaluate the performance of prediction intervals in forecasting models. They assess the coverage of the true values by the prediction intervals, but they focus on different aspects:

- **Unconditional coverage**: Unconditional coverage assesses the proportion of true values that fall within the prediction intervals without considering specific conditions or patterns. It measures the overall ability of the prediction intervals to capture the true values across the entire dataset. A model with good unconditional coverage will include the true values within the prediction intervals for a specified proportion (e.g., 95%) of the time. Unconditional coverage is useful for evaluating the general performance of a model, but it does not account for potential dependencies between observations or changes in data distribution.

- **Conditional coverage**: On the other hand, conditional coverage evaluates the performance of prediction intervals while accounting for specific conditions or patterns in the data. It examines how well the prediction intervals capture the true values when considering subsets of the data that share certain characteristics or dependencies (e.g., time periods, categories, etc.). A model with good conditional coverage will maintain the desired coverage rate for each specific condition or subset of the data. Conditional coverage provides a more nuanced evaluation of a model's performance, helping to identify potential weaknesses or biases in the model's predictions for certain data subsets.

In summary, unconditional coverage evaluates the overall ability of a model's prediction intervals to include the true values. In contrast, conditional coverage assesses the performance of prediction intervals within specific conditions or data subsets. Both criteria are important for understanding the performance of forecasting models, but they focus on different aspects of a model's predictions.

Conformal prediction is a framework for producing reliable and valid predictions with quantifiable uncertainty. It can be applied to a wide range of machine learning, statistical, or deep learning models and other prediction methods. Here, we'll discuss the relationship of conformal prediction with other frameworks:

- **Traditional machine learning frameworks**: Conformal prediction can be combined with traditional machine learning methods (e.g., linear regression, SVM, decision trees, etc.) to provide valid confidence or credibility measures for the predictions. By doing so, conformal prediction enhances these methods, giving users a better understanding of the uncertainty associated with each prediction.

- **Ensemble methods**: Ensemble methods such as bagging, boosting, and random forests can also benefit from conformal prediction. By adding conformal prediction to these methods, the ensemble can produce a point estimate and a prediction interval or set with associated confidence levels.

- **Deep learning frameworks**: Conformal prediction can be integrated with deep learning models, such as neural networks, to provide quantifiable uncertainty estimates for their predictions. This allows practitioners to better understand the reliability of the predictions produced by these complex models.

- **Bayesian frameworks**: Bayesian methods inherently provide uncertainty quantification through probability distributions. However, conformal prediction can still be combined with Bayesian frameworks to offer a frequentist approach to uncertainty quantification. This combination can provide a complementary perspective on the uncertainty associated with predictions.

- **Model validation techniques**: Conformal prediction can be used alongside model validation techniques such as cross-validation or bootstrapping to assess the performance of a model. While these validation techniques evaluate the model's accuracy and generalization, conformal prediction provides a complementary perspective on the model's uncertainty quantification.

In *Chapter 4*, we will look into the concepts of validity and efficiency in the context of probabilistic prediction models, building upon the foundations laid in the previous chapters.

Summary

In this chapter, we have deep-dived into the fundamentals and mathematical foundations of conformal prediction, a powerful and versatile probabilistic prediction framework. We have learned about different measures of nonconformity used in classification and regression, building solid foundations for applying conformal prediction to your industry applications.

In the next chapter, we'll cover the concepts of validity and efficiency in the context of probabilistic prediction models, building upon the foundations laid in the previous chapters.

4

Validity and Efficiency of Conformal Prediction

In this chapter, we will dive deeper into the concepts of validity and efficiency in the context of probabilistic prediction models, building upon the foundations laid in the previous chapters.

Validity and efficiency are crucial aspects that ensure the practicality and robustness of prediction models across a wide range of industry applications. Understanding these concepts and their implications will enable you to develop unbiased and high-performing models that can reliably support decision-making and risk assessment processes.

In this chapter, we will explore the definitions, metrics, and examples of valid and efficient models and discuss the automatic validity guarantees provided by **conformal prediction**, a cutting-edge approach to uncertainty quantification. By the end of this chapter, you will be equipped with the knowledge necessary to assess and improve the validity and efficiency of your predictive models, paving the way for more reliable and effective applications in your respective fields.

In this chapter, we will cover the following topics:

- The validity of probabilistic predictors
- The efficiency of probabilistic predictors

The validity of probabilistic predictors

We start by summarizing the reasons why unbiased point prediction models are important across various domains and applications:

- **Accuracy and reliability**: An unbiased model ensures that the predictions it generates are accurate and reliable on average, meaning that the model is neither systematically overestimating nor underestimating the true values. This accuracy is crucial for making well-informed decisions, minimizing risks, and improving the overall performance of a system.

- **Trust and credibility**: Unbiased prediction models help build trust and credibility among stakeholders, as they provide a reliable basis for decision-making. Users can have more confidence in the outputs generated by an unbiased model, knowing that it is not skewed or favoring any specific outcome.

- **Fairness and equity**: In some applications, such as finance, healthcare, and human resources, unbiased models are essential to ensure fairness and equity among different groups or individuals. Biased models can inadvertently reinforce existing inequalities or create new ones, leading to unfair treatment or allocation of resources.

- **Generalizability**: Unbiased models are more likely to generalize well to new, unseen data because they accurately represent the underlying relationships in the data. In contrast, biased models may perform poorly when applied to new data or different conditions, leading to unexpected errors or suboptimal outcomes.

- **Regulatory compliance**: In certain industries, unbiased models are a requirement for regulatory compliance. For example, in finance, healthcare, and insurance, models must be free of bias to meet regulatory standards and ensure that customers are treated fairly and risks are managed effectively.

In the context of conformal prediction, validity refers to the ability of a prediction model to provide accurate, reliable, and unbiased estimates of the uncertainty associated with its predictions. More specifically, a valid conformal predictor generates prediction intervals that contain the true values of the target variable with a predefined coverage probability, ensuring that the model's uncertainty quantification is reliable and well calibrated.

The significance of validity in conformal prediction can be understood from various perspectives:

- **Confidence in predictions**: Valid conformal predictors allow users to have confidence in the prediction intervals they generate, as they know that these intervals truly reflect the uncertainty in the predictions. For instance, if a conformal predictor produces a 95% prediction interval for a certain data point, users can trust that there is a 95% probability that the true value lies within that interval. This confidence is crucial for decision-making and risk management in various applications.

- **Robustness to model misspecification**: One of the key strengths of conformal prediction is its ability to provide valid uncertainty estimates even when the underlying prediction model is misspecified or imperfect. This robustness to model misspecification is particularly valuable in real-world settings, where the true data-generating process is often unknown or complex and the available models may only provide approximations of the underlying relationships.

- **Non-parametric nature**: Conformal prediction is a non-parametric method, meaning that it does not rely on any specific assumptions about the data distribution or prediction errors. This non-parametric property further contributes to the validity of conformal predictors, as they can adapt to different data structures and provide accurate uncertainty estimates without requiring explicit knowledge of the underlying distributions.

- **Applicability across domains**: The validity of conformal prediction is a universal property that holds across various domains and applications. This universality allows practitioners to leverage conformal prediction in diverse fields, such as finance, healthcare, energy, and transportation, knowing that the uncertainty estimates provided by conformal predictors will be valid and reliable regardless of the specific context.

- **Automatic validity guarantees**: A key advantage of conformal prediction over traditional methods is its ability to provide automatic validity guarantees, meaning that the uncertainty estimates it produces are guaranteed to be valid under mild assumptions, such as the exchangeability of the data. This automatic validity ensures that conformal predictors maintain their reliability even as new data points are added or as the underlying relationships evolve over time.

In conformal prediction, validity is mathematically defined in terms of the coverage probability of the prediction intervals or regions generated by the conformal predictor. A conformal predictor is valid if, for any desired confidence level $(1-\alpha)$, the proportion of true target values contained within their corresponding prediction intervals is at least $(1-\alpha)$, on average, across multiple instances.

Mathematically, let's denote the target variable as Y and the prediction interval (in regression) or set (in classification) as $I(x, \alpha)$, where x represents the features of the test data point and α is the significance level $(\alpha \in [0, 1])$. The conformal predictor is valid if the following condition holds:

$$P(Y \in I(x, \alpha)) \geq 1 - \alpha$$

This condition states that the probability that the true target value Y is within the prediction interval $I(x, \alpha)$ is at least $(1-\alpha)$ for any given input data point x.

Validity in conformal prediction is closely related to the calibration concept in probabilistic prediction. A calibrated predictor generates prediction intervals with the correct coverage probability, ensuring that the uncertainty estimates it provides are well aligned with the true underlying uncertainty in the data.

It is important to note that validity in conformal prediction is guaranteed under the assumption of exchangeability, which requires that the observed data points are exchangeable with future, unseen data points. This assumption holds for **independent and identically distributed** (IID) data. In addition, successful modifications of conformal prediction have been developed to address non-exchangeable data, including many successful models for time series.

Classifier calibration

Classifier calibration ensures that the predicted probabilities of an event match the true probabilities or frequencies of that event occurring. For example, in weather forecasting, calibration ensures that the forecasted probability of rain aligns with the actual occurrence of rain over a series of predictions.

The concept of classifier calibration has been applied to weather forecasting since the 1950s, pioneered by Glen Brier. In the case of rain forecasting, a forecaster might declare an 80% chance of rain. If, on average, rain occurs 60% of the time following such statements, we consider the forecast well calibrated.

Let's consider a weather forecaster who makes a statement that there is an x% chance of rain for a particular day. To assess the calibration of the forecaster's predictions, we would collect a series of similar predictions along with their corresponding outcomes (whether it rained or not). For example, suppose we gather 100 instances in which the forecaster predicted a 60% chance of rain. If the forecaster's predictions are well calibrated, it should rain on approximately 60 of those 100 days, resulting in an observed frequency of rain that matches the predicted probability.

But what would this mean in practice? Let's consider an example to understand this concept better.

Suppose, over time, a weather forecaster made 10 predictions of an x% chance of rain. In the following table, we show these forecasts and the actual outcome:

Day	Forecasted probability of rain	Actual outcome (rain)
1	80%	Yes
2	60%	No
3	90%	Yes
4	30%	No
5	70%	Yes
6	50%	No
7	80%	Yes
8	20%	No
9	40%	Yes
10	60%	Yes

Table 4.1 – Forecasted probability of rain versus actual outcome

To determine whether this forecast is well calibrated, we need to compare the forecasted rain probabilities with the actual outcomes for each probability level. We can group the forecasts by their probability levels and calculate the observed frequency of rain for each group.

Chance of rain	20%	30%	40%	50%	60%	70%	80%	90%
Predicted	1 day	1 day	1 day	1 day	2 days	1 day	2 days	1 day
Rained	0 days	0 days	1 day	0 days	1 day	1 day	2 days	1 day
Observed frequency	0%	0%	100%	0%	50%	100%	100%	100%

Table 4.2 – Forecasted probability of rain versus observed frequency

Based on the observed frequencies, we can see that the forecast is not well calibrated. The observed frequencies do not align with the forecasted probabilities for most probability levels. For example, on the two days with a 60% chance of rain, it only rained once (50% of the predicted frequency), and on the day with a 50% chance of rain, it didn't rain at all (0% the predicted frequency).

How can we aggregate these results into certain metrics? We could use the **Brier score**, which we encountered in previous chapters. The Brier score is a commonly used calibration metric for binary classification problems.

Recall that the Brier score is calculated as the mean squared difference between the predicted probabilities and the true binary outcomes (0 or 1):

$$Brier\ score = (1/N)\ \Sigma(prediction\_i - outcome\_i)^{\wedge}2$$

where N is the number of predictions, *prediction_i* is the predicted probability for the i-th instance, and *outcome_i* is the true binary outcome for the i-th instance (1 for rain, 0 for no rain).

Now we can compute the Brier score:

$Brier\ score = (1/10) * (0.04 + 0.36 + 0.01 + 0.09 + 0.09 + 0.25 + 0.04 + 0.04 + 0.36 + 0.16) = 0.144$

A lower Brier score indicates better model performance and, consequently, better calibration. However, with a reference or comparison to other models, it is easier to determine whether a Brier score of 0.144 is good or not. Additionally, it is important to remember that this assessment is based on a limited sample size of only 10 days, which may not provide an accurate representation of the forecast's calibration over a longer period.

We can also create a calibration diagram, also known as a reliability diagram, to assess the calibration of probabilistic prediction models. The diagram plots the predicted probabilities (grouped into bins) on the x axis against the observed frequencies of the event on the y axis. A well-calibrated model would have points along the diagonal (45-degree angle), indicating that the predicted probabilities match the observed frequencies.

As we can see from *Table 4.2*, the forecast is not well calibrated. The observed frequencies do not match the forecasted probabilities for most probability levels.

There are two types of miscalibration: *underconfidence* and *overconfidence*. When a classifier exhibits underconfidence, it underestimates its ability to distinguish between classes, performing better in practice than its predictions suggest. In contrast, an overconfident classifier overestimates its capacity to separate classes, performing worse than its predicted probabilities imply.

Another metric that can be used to evaluate calibration is log loss, also known as logarithmic loss or cross-entropy, which is a metric used to evaluate the performance of classification models that produce probability estimates for each class. It measures the divergence between the true and predicted probability distributions, penalizing incorrect and uncertain predictions.

The concept of log loss is based on the idea that a classifier should not only predict the correct class but also be confident in its prediction. Log loss quantifies the uncertainty in the predicted probabilities by comparing them with the actual outcomes.

For binary classification, log loss is defined as follows:

$$Log\ loss = -\left(y * log(p) + (1 - y) * log(1 - p)\right)$$

In this scenario, y stands for the true class label, which can be either 0 or 1. The symbol p signifies the predicted probability for the positive class (class 1). The term log indicates the natural logarithm. The log loss is calculated for each instance and then averaged across all instances to obtain the final log loss value.

In a calibration context, log loss can be used to assess how well the predicted probabilities match the true outcomes. A well-calibrated model will have a lower log loss, as the predicted probabilities will be closer to the actual class labels. Conversely, a poorly calibrated model will have a higher log loss, indicating a discrepancy between the predicted probabilities and the true outcomes.

It is important to note that log loss alone might not be sufficient to assess calibration, as it also depends on the classification accuracy. However, when used in conjunction with other metrics, such as calibration diagrams, log loss can provide valuable insights into the calibration of a classification model. In practice, log loss is often used alongside the Brier score to evaluate a model's calibration. When both metrics agree on the relative calibration of two models, this provides stronger evidence than relying on a single calibration metric, such as log loss or Brier loss, alone. A more comprehensive assessment of a model's calibration can be achieved by considering both metrics.

Recall the rain prediction example from earlier.

To calculate the log loss for this example, we will use this formula:

$$Log\ loss = -\left(y * log(p) + (1 - y) * log(1 - p)\right)$$

Here, y is the true class label (0 or 1) and p is the predicted probability of the positive class (rain).

Let's compute the log loss for each day:

- Day 1: (1 * log(0.8) + (1 - 1) * log(1 - 0.8)) = 0.223
- Day 2: (0 * log(0.8) + (1 - 0) * log(1 - 0.8)) = 1.609
- Day 3: (1 * log(0.6) + (1 - 1) * log(1 - 0.6)) = 0.511
- Day 4: (1 * log(0.7) + (1 - 1) * log(1 - 0.7)) = 0.357
- Day 5: (1 * log(0.9) + (1 - 1) * log(1 - 0.9)) = 0.105
- Day 6: (0 * log(0.7) + (1 - 0) * log(1 - 0.7)) = 1.204
- Day 7: (1 * log(0.6) + (1 - 1) * log(1 - 0.6)) = 0.511

- Day 8: $(1 * \log(0.5) + (1 - 1) * \log(1 - 0.5)) = 0.693$
- Day 9: $(0 * \log(0.6) + (1 - 0) * \log(1 - 0.6)) = 0.916$
- Day 10: $(0 * \log(0.4) + (1 - 0) * \log(1 - 0.4)) = 0.511$

Now, we can compute the average log loss across all 10 days:

Average log loss $= (0.223 + 1.609 + 0.511 + 0.357 + 0.105 + 1.204 + 0.511 + 0.693 + 0.916 + 0.511) / 10 \approx 0.664$

The average log loss for this rain prediction example is approximately 0.664.

A question naturally arises: of statistical, machine, and deep learning, which are well calibrated and which are not?

As a general guideline, it is important to remember that most machine learning models are miscalibrated to some extent, with varying degrees of severity. However, logistic regression has its own limitations and may only be suitable for some applications due to its relatively simpler modeling capacity.

A classical study of calibration is the paper *Predicting Good Probabilities With Supervised Learning (2005):* `https://www.cs.cornell.edu/~alexn/papers/calibration.icml05.crc.rev3.pdf` which examined the calibration properties of various supervised classification algorithms. The paper found that maximum margins, such as support vector machines and boosted trees, produced miscalibrated class scores and tended to push predictions close to 0 and 1, while other methods, such as naïve Bayes, pushed predictions in the other directions.

While it was initially thought that simple neural networks produced calibrated predictions, this conclusion has since been reevaluated. In a more recent paper titled *Are Traditional Neural Networks Well-Calibrated?* (`https://ieeexplore.ieee.org/document/8851962`), the authors showed that individual multilayer perceptrons, as well as ensembles of multilayer perceptrons, frequently display poor calibration.

The efficiency of probabilistic predictors

Efficiency is a performance metric used to evaluate probabilistic predictors. It measures how precise or informative the prediction intervals or regions are. In other words, efficiency indicates how tight or narrow the predicted probability distributions are. Smaller intervals or regions are considered more efficient, as they convey more certainty about the predicted outcomes.

While validity focuses on ensuring that the error rate is controlled, efficiency assesses the usefulness and precision of the predictions. An efficient predictor provides more specific information about the possible outcomes, whereas a less efficient predictor generates wider intervals or regions, resulting in less precise information.

There is an inherent trade-off between validity and efficiency. A conformal predictor can always achieve perfect validity by outputting very wide prediction sets that encompass all possible outcomes. However, this lacks efficiency, as the predictions are too conservative and imprecise.

On the other hand, a model can output very narrow, precise predictions but may fail on the validity criteria by making erroneous predictions more than the allowed threshold. This results from overconfidence and unreliable probability estimates.

Ideally, a conformal predictor finds an optimal balance; the predictions are as tight as possible while still meeting the validity guarantee. This ensures accuracy and precision without being overly conservative or exceeding the error rate threshold.

In conformal prediction, efficiency is typically measured by evaluating the size of the prediction intervals or regions generated by the conformal predictor. Smaller intervals or regions are considered more efficient, as they provide more precise information about the possible outcomes. Here are a few common ways to measure efficiency in conformal prediction:

- **Prediction interval length**: For regression problems, the average length of the prediction intervals can be calculated by finding the difference between the upper and lower bounds of each interval and then averaging these differences across all instances. Smaller average lengths indicate higher efficiency.

- **Prediction set size**: For classification problems, the size of the prediction sets can be evaluated. A smaller prediction set contains fewer class labels and is considered more efficient. One way to measure this is by computing the average size of the prediction sets across all instances. A lower average set size indicates better efficiency.

- **Coverage probability**: Coverage probability measures the proportion of true outcomes that fall within the prediction intervals or regions. While it is mainly used to evaluate the validity of conformal predictors, it can also provide insights into efficiency. A predictor with tight intervals or regions will have a higher coverage probability, indicating better efficiency.

- **P-value histograms**: In conformal prediction, p-values are calculated for each instance and class label. Examining the distribution of p-values can provide insights into efficiency. A uniform distribution of p-values suggests that the predictor is valid but not necessarily efficient, while a more concentrated distribution (e.g., with p-values close to 0 or 1) implies greater efficiency.

We have already seen in the previous chapters how conformal prediction guarantees the automatic validity of prediction sets by constructing prediction intervals (for regression) or prediction sets (for classification) that come with a guaranteed error rate, which is determined by a user-defined confidence level. The key idea behind conformal prediction is to use past data and the observed behavior of a given machine learning model to estimate the uncertainty in its predictions.

Let's recap the stages in **inductive conformal prediction,** which consists of two main phases: the calibration phase and the prediction phase. Here's an outline of how it works:

- **Calibration phase**: In this phase, a machine learning model is trained on a dataset, and a nonconformity measure is calculated for each instance in the dataset. The nonconformity measure quantifies the strangeness or atypicality of an instance with respect to the rest of the data.

- **Prediction phase**: When a new instance needs to be predicted, the nonconformity measure, for instance, is calculated using the same nonconformity measure function used in the calibration phase. The instance's nonconformity score is then compared to the nonconformity scores of the calibration instances. A p-value is computed for each possible outcome, reflecting the proportion of calibration instances with nonconformity scores higher than or equal to that of the new instance. The p-values can be interpreted as a measure of how likely it is for the instance to belong to each class (for classification) or to fall within a certain range (for regression).

- **Prediction intervals or sets**: Based on the computed p-values and the user-defined confidence level, prediction intervals (for regression) or prediction sets (for classification) are constructed. These intervals or sets are guaranteed to contain the true outcome with a probability equal to the chosen confidence level. For instance, if the confidence level is set to 0.95, the true outcome will fall within the prediction interval or set 95% of the time.

By ensuring that the prediction intervals or sets contain the true outcomes with the desired probability, conformal prediction provides automatic validity for predictions. It is worth noting that while conformal prediction guarantees validity, it does not necessarily guarantee efficiency, which depends on the precision of the prediction intervals or sets.

In *Chapter 4*, we will look at and learn about different types of conformal prediction with practical examples.

Summary

In this chapter, we have deep-dived into the concepts of validity and efficiency in the context of probabilistic prediction models, building upon the foundations laid in the previous chapters. We have looked at the definitions of validity and efficiency and learned about various metrics that can be used to evaluate and compare different models in terms of validity and efficiency.

In the next chapter, we will learn about different families of conformal predictors and explore various approaches to quantifying uncertainty.

Types of Conformal Predictors

This chapter describes different families of conformal predictors, exploring various approaches to quantifying uncertainty. Through practical examples, we provide an intermediate-level understanding of these techniques and how they can be applied to real-world situations.

Here are examples of how companies are using conformal prediction.

At a high-profile AI developer conference called GTC 2023 (https://www.nvidia.com/gtc/), Bill Dally, NVIDIA's chief scientist and SVP of research, offered insights into one of NVIDIA's R&D primary focuses, which is in conformal prediction (https://www.hpcwire.com/2023/03/28/whats-stirring-in-nvidias-rd-lab-chief-scientist-bill-dally-provides-a-peek/).

Traditional machine learning models for autonomous vehicles output a single classification (e.g., pedestrian or no pedestrian on the road) and position estimate for detected objects. However, NVIDIA wants to produce a set of potential outputs with probabilities; for example, an object could be a pedestrian (80% probability) or cyclist (20% probability) at a position of 20 +/- 1 meters.

This allows the vehicle's planner to guarantee safe actions accounting for multiple possible outcomes. Rather than just the most likely label and position, conformal prediction provides a range of plausible options, such as "pedestrian at 19–21 meters" with 80% confidence.

NVIDIA uses a nonconformity function to calculate probabilities that measure how strange or different each potential label and position is compared to the training data. This generates a multi-modal predictive distribution reflecting uncertainty.

Conformal prediction gives NVIDIA's vehicles a reliable way to quantify uncertainty and consider multiple interpretations of the environment. By planning for the entire set of plausible outcomes rather than just the single most likely one, conformal prediction improves robustness and safety.

In the realm of machine learning, quantifying uncertainty and providing reliable predictions is of significant importance. Conformal prediction is an innovative technique that allows us to construct prediction sets (in classification) and prediction intervals (in regression), offering a measure of confidence in our predictions.

This chapter aims to provide a deeper understanding of the different types of conformal predictors and their respective approaches to quantifying uncertainty. Through practical examples, we will illustrate how these techniques can be applied to various machine learning tasks.

In this chapter, we will explore the following topics:

- Foundations of classical and inductive conformal predictors

- Examining algorithmic descriptions of conformal predictors

- Mathematical formulations and practical examples

- Advantages and limitations of conformal predictors

- Guidelines for choosing the most suitable conformal predictor for specific problem domains

Let's start with the classical conformal predictors.

Understanding classical predictors

Before we deep dive into the intricacies of conformal predictors, let's briefly recap the key concepts from the previous chapters. Conformal prediction is a framework that enables creating confidence regions for our predictions while controlling the error rate.

This approach is especially beneficial in situations where a measure of uncertainty is essential, such as in medical diagnosis, self-driving cars, or financial risk management. The framework encompasses two main types of conformal predictors: **classical** and **inductive**.

Classical transductive conformal prediction (TCP) is the original form of conformal prediction developed by the inventors of Conformal prediction. It forms the basis for understanding the general principles of conformal predictors. Classical Conformal prediction was developed to construct prediction regions that conform to a specified confidence level. The critical aspect of classical Conformal prediction is its distribution-free nature, meaning it makes no assumptions about the data distribution. Thus, it can be applied to any machine learning algorithm, making it algorithm-agnostic.

In contrast to the widely used inductive conformal prediction, classical TCP does not require a separate calibration set, enabling a more efficient utilization of the entire dataset.

As a result, for smaller datasets, it can generate more accurate predictions. This approach allows statistical, machine learning, and deep learning models to fully capitalize on all available data, potentially leading to more efficient (narrower) prediction sets and intervals.

Let's discuss how classical TCP can be used in classification problems.

Applying TCP for classification problems

In classification tasks, not only do we seek to assign labels, but we aim to do so confidently and accurately. This is where classical TCP shines. As we delve into its use in classification, we will cover its unique approach and advantages over traditional techniques. Ready to explore? Let's dive into the nuances of TCP for classification!

In the previous chapters, we discussed the core concept of the conformal prediction framework, which involves assigning a nonconformity measure (or strangeness) to each object in the dataset. This measure is utilized to rank the objects within the dataset. Subsequently, when predicting a new object, a prediction region is created that encompasses values linked to a specific proportion of the dataset objects, determined by their strangeness scores. This proportion corresponds to the desired confidence level for the predictions.

In contrast to the more popular inductive conformal prediction method, which relies on a calibration set to rank objects based on their nonconformity scores, classical TCP employs the entire dataset in conjunction with the features of the new object to establish the prediction region.

While this approach can be computationally expensive, it allows you to fully leverage the whole dataset to capture changes in the data distribution, providing more accurate prediction regions.

However, classical conformal prediction has some limitations, too. For instance, it may not be feasible for large datasets or real-time applications because it requires retraining the underlying point prediction model for each new prediction.

Classical conformal prediction is a process that involves several key steps. Here, we outline these steps to provide a clear understanding of the procedure:

1. **Dataset preparation**: Divide the dataset into training and test sets. The training set is used to train the machine learning model, while the test set is used to evaluate the performance of the conformal predictor.

2. **Model training**: Train the underlying machine learning model using the training dataset. This point prediction model will generate point predictions for new objects.

3. **Nonconformity measure calculation**: Define a nonconformity (strangeness) measure that quantifies how different an object is from the other objects in the dataset. For each object in the training dataset, calculate its nonconformity score using the trained model.

4. **New object nonconformity score**: When a new object (without its label) is introduced, calculate its nonconformity score using the same nonconformity measure and the trained point prediction model.

5. **Ranking**: Based on calculated nonconformity scores, rank all objects, including the objects in the training dataset and the new object.

6. **Prediction region**: Determine the desired confidence level for the prediction. Identify the proportion of objects in the ranked set corresponding to this confidence level. Form a prediction set that includes the values associated with these objects.

Let's clarify these concepts using a practical example with the hinge loss nonconformity measure, which we discussed in the previous chapters. As a reminder, hinge loss (also known as inverse probability or

LAC loss) is a nonconformity measure calculated as *1-P(y|x)*, where *P(y|x)* represents the class score produced by the underlying model for the actual class.

The hinge loss nonconformity measure intuitively measures the difference between the probability score generated by an ideal classifier for the correct class (which should ideally be 1) and the classification score produced by the classifier model. It quantifies how far the model's prediction is from the perfect classification, with larger nonconformity scores indicating a more significant discrepancy between the ideal and actual predictions.

To compute the inverse probability (hinge) nonconformity score, consider an example where your classifier generates two scores: *class_0 = 0.6* and *class_1 = 0.4,* with the actual label *y=1*. To determine the nonconformity score, subtract the probability of the true class (in this case, 0.4 from 1).

The resulting inverse probability (hinge) nonconformity score is 0.6.

The hinge (inverse probability) score is lower when the underlying machine learning classification model performs better. This performance is influenced by a range of factors, such as the size and complexity of the dataset, the type of machine learning model employed, and the quality of the model's construction. In other words, a well-built model using an appropriate machine learning technique for the given dataset will generally yield lower hinge scores, indicating better predictions.

The training process is the critical difference between TCP and **inductive conformal prediction (ICP)**. In ICP, the underlying classifier is trained only once on the training set and the calibration of the conformal prediction model happens on the calibration dataset.

In contrast, with TCP, the classifier is trained by appending each test point to the training set twice, each time assigning potential labels 0 and 1. This procedure is repeated for every point in the test set. As a result, the underlying classifier model is trained *2 x m* times, where m is the number of points in your test set. This may become computationally expensive for large datasets, and for such datasets, using ICP might be a more suitable choice.

However, the computational cost is typically manageable for medium and small datasets. To obtain potentially better point predictions and narrower probability intervals, you might consider TCP, which achieves better prediction intervals by training the classifier model *2 x m* times. Many algorithms, such as logistic regression, are fast and well suited for this approach.

The overall methodology for training TCP remains fundamentally unchanged. The TCP algorithm process is as follows:

1. Train the underlying classifier on the entire training set.
2. Append each test point to the training set with each possible class label one class label at a time.
3. For each appended test point with a postulated label, retrain the classifier and compute the nonconformity score for the test point given the postulated label.
4. Calculate the p-values for each postulated label, comparing the test point's nonconformity score to the scores of the points in the training set.

5. For each test point and each postulated label, include the postulated label in the prediction set if its p-value is greater than or equal to the chosen significance level.

We will illustrate the TCP approach with a practical classification task example using the German credit dataset (`https://www.openml.org/d/31`), a classical dataset describing good and bad credit risk based on features such as loan duration, credit history, employment, property, age, housing, and so on.

In the GitHub repo for the book, you will find a notebook (`https://github.com/PacktPublishing/Practical-Guide-to-Applied-Conformal-Prediction/blob/main/Chapter_05_TCP.ipynb`) describing how TCP works that you can work through to understand the key concepts of TCP in practice.

checking_status	duration	credit_history	purpose	credit_amount	savings_status	employment	installment_commitment	personal_status
828	0.0	36.0	2.0	1.0	8335.0	4.0	4.0	3.0	2.0
997	3.0	12.0	2.0	3.0	804.0	0.0	4.0	4.0	2.0
148	0.0	36.0	4.0	2.0	5371.0	0.0	2.0	3.0	2.0
735	1.0	36.0	1.0	4.0	3990.0	4.0	1.0	3.0	1.0
130	1.0	48.0	2.0	0.0	8487.0	4.0	3.0	1.0	1.0
...	...	...	...	...	...	...	...	...	...
545	0.0	24.0	3.0	0.0	1333.0	0.0	0.0	4.0	2.0
298	3.0	18.0	2.0	2.0	2515.0	0.0	2.0	3.0	2.0
417	0.0	18.0	3.0	6.0	8471.0	4.0	2.0	1.0	1.0
749	3.0	15.0	2.0	1.0	3029.0	0.0	3.0	2.0	2.0
30	1.0	18.0	2.0	9.0	1913.0	3.0	1.0	3.0	3.0

801 rows × 20 columns

Figure 5.1 – German credit dataset

For clarity, let's examine the first test point with the original index 30, which has now been appended to the end of the training set. We will use this extended training set we created to train the classical transductive conformal predictor. This new dataset created using the code in the notebook incorporates all points from the original training set and the single test point.

We now have a feature set to train two classification models: one with an assumed label of the test point of 0 and another with an assumed label of 1. We train two models using any classifier (in this case, *Logistic Regression* from scikit-learn) and calculate nonconformity scores using the described procedure.

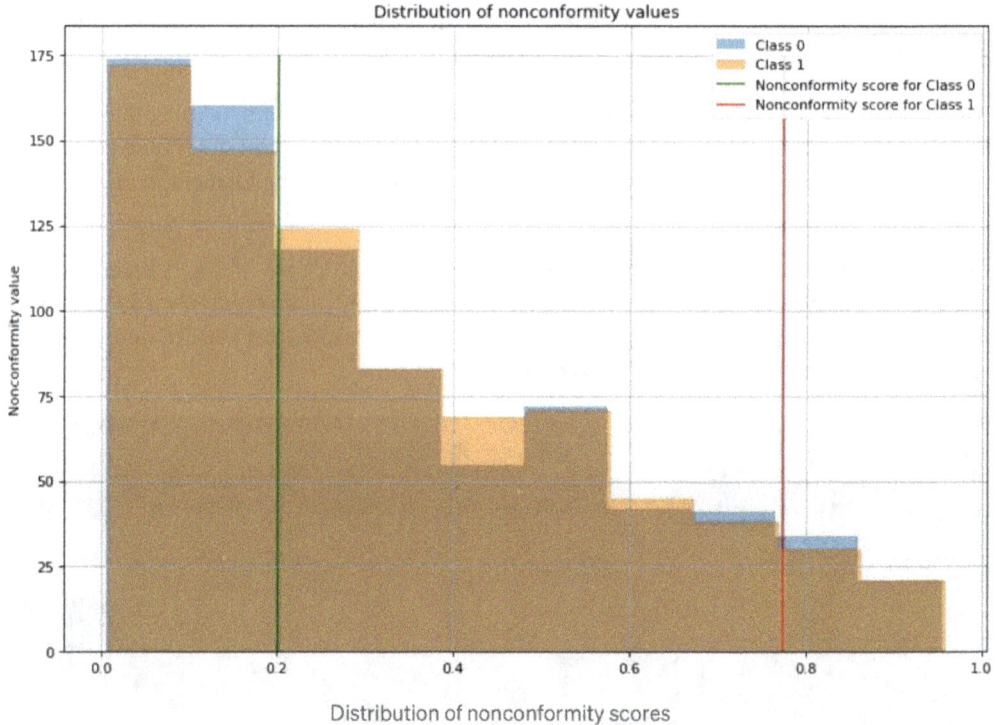

Figure 5.2 – Distribution of nonconformity scores

From the distribution of nonconformity scores, we observe that the nonconformity score for *label 0* (represented by the green vertical line) is relatively typical (more conforming) to the training set. In contrast, the nonconformity score for the potential *label 1* (represented by the red vertical line) is in a low-density probability area.

This suggests that the test object is likelier to be assigned a label of 0, while label 1 is less probable. However, conformal prediction is a robust mathematical machine learning framework, so we must quantify and statistically test this decision. This is where p-values come into play.

Let's take a moment to revisit the conventional process of calculating p-values, which we previously explored in *Chapter 3*, using the formula from Vovk's book *Algorithmic Learning in a Random World*.

p-values can be computed as follows:

$$p = \left(|z_i : \alpha_i \geq : \alpha_T| + 1\right)/(n + 1)$$

Here, the nonconformity score of a new test point is compared with the nonconformity scores of points in the training set. Essentially, the nonconformity score quantifies the *strangeness* or novelty of the new test object compared to the previously encountered objects in the training dataset.

According to the formula, what we need to do is to check (for each test point and each potential value of label 0 and 1) how many objects in the set of training data appended with the test point using the postulated label have nonconformity values that are larger or equal to the nonconformity score of the test point.

We then divide it by the number of training points *(n+1)* (+1 accounts for the test point that we appended to the training set). As a result, we obtain two p-values for each test point: one for class 0 and one for class 1.

The central concept of conformal prediction revolves around utilizing nonconformity values for each test point to evaluate how well it aligns with the training set. By computing p-values based on this evaluation, we can conduct robust statistical tests to determine if each potential label value should be included in the prediction set.

Let's say we have a postulated label (either 0 or 1). If there are sufficient instances in the training set with nonconformity values equal to or greater than the test point's nonconformity value, then we infer that this postulated label aligns well with the observed data. As a result, we incorporate this label into our prediction set. Conversely, if the postulated label does not correspond well with the observed data, we refrain from including it in the prediction set.

In essence, this procedure echoes the principles of statistical hypothesis testing. For each hypothesized label value, we establish a null hypothesis, asserting that the label could be part of the prediction set if its associated p-value exceeds a pre-defined significance level. If the p-value falls short of this threshold, we discard the null hypothesis. This implies that the proposed label doesn't adequately match the pattern found in the training data, leading us to exclude it from our prediction set.

For example, let's say we have calculated two p-values for the first test object:

1. Assume for label 0 that the p-value is 0.55. Since the p-value is larger than the significance level (0.05), we include the hypothesized label (0 in this case) in the prediction set for this test point.

2. Now assume for label 1 that the p-value is 0.002. Since the p-value is smaller than the significance level (0.05), we cannot include the hypothesized label (1 in this case) in the prediction set for this test point.

3. Thus, the final prediction set for this point is 0.

In the context of TCP, the key distinction between binary and multiclass classification lies in the number of potential labels taken into account for each test point. In a binary classification scenario, only two labels exist (0 and 1). In contrast, multiclass classification involves a greater number of classes (for instance, C1, C2, C3, ...).

The main difference from the binary classification is that we will have to repeat the process for each possible class label, increasing computational complexity as you need to retrain the classifier for each test point and each potential label. However, the overall method for obtaining the prediction set remains the same.

After delving into the nuances of TCP for classification, let's pivot our focus. Next up, we'll explore the intricacies of employing TCP in regression contexts. This approach offers unique challenges and benefits, so let's dive in!

Applying TCP for regression problems

TCP can also be applied to regression problems. The process for TCP in regression is similar to the one used for classification, with some differences in computing nonconformity scores and prediction intervals. Here is an algorithmic description of the TCP for regression problems:

1. Train the underlying regression model on the entire training set.

2. For each test point, create a grid of potential target values. The granularity of this grid depends on the desired precision and the problem's nature.

3. For each test point and each potential target value on the grid, append the test point to the training set with the associated target value.

4. For each appended test point with a postulated target value, retrain the regression model and compute the nonconformity score for the given postulated target value. The nonconformity score can be computed as the absolute difference between the predicted value and the true value of the appended point, or you can calculate it by using other error metrics such as **mean squared error (MSE)**.

5. Calculate the p-values for each postulated target value by comparing the test point's nonconformity score for each value on the grid of potential target values to the scores of the points in the training set.

6. For each test point, include the postulated target value in the prediction interval if its p-value is greater than or equal to the chosen significance level.

The prediction set for a regression problem will be an interval rather than a set of discrete labels, as in classification. The main difference from the classification is that you will have to repeat the process for each potential target value in the grid, which could increase computational complexity. However, the overall method for obtaining the prediction interval remains the same.

We conclude the section about TCP by summarizing the advantages and limitations of TCP.

Advantages

TCP has several advantages over alternative methods of uncertainty quantification:

- **Distribution-free**: Transductive conformal predictors do not make any assumptions about the distribution of the data, making them suitable for various types of data

- **Validity**: They provide prediction intervals with a guaranteed coverage probability, allowing for a reliable measure of uncertainty

- **Adaptability**: Conformal predictors can be applied to various machine learning models, making them versatile and easily adaptable to different settings

- **Better prediction intervals**: Transductive conformal predictors generally produce more precise prediction intervals compared to inductive conformal predictors since they fully utilize the dataset for training the underlying point prediction model

But there are a few limitations as well:

- **Computational expense**: TCP requires retraining the model for each test point and for each potential class label (in classification) or each potential target value on the grid regression, making it computationally expensive, particularly for large datasets

- **Not ideal for online learning**: Due to the computational expense, transductive conformal predictors are not well suited for online learning scenarios where models must be continuously updated

- **Complexity**: Implementing transductive conformal predictors can be more complicated than traditional machine learning models, potentially posing a barrier to widespread adoption

Transductive conformal predictors offer several advantages in providing reliable, distribution-free prediction intervals. However, their computational expense and scalability limitations should be considered, particularly for large-scale or online learning applications.

Building upon our exploration of TCP, it's time to turn our attention to another intriguing variant: inductive conformal predictors. Differing from its classical counterpart in key ways, this approach brings a new set of strategies and benefits to the table. Ready to delve into the mechanics and merits of inductive conformal predictors? Let's embark on this enlightening journey!

Understanding inductive conformal predictors

ICP is a variant of conformal prediction that provides valid predictive regions under the same assumptions as classical conformal prediction and has the added benefit of improved computational efficiency, which is particularly useful when dealing with large datasets.

ICPs present a highly efficient and effective solution within the realm of machine learning. They provide a form of conformal prediction that caters to larger datasets, making it highly suitable for real-world applications that involve extensive data volumes. ICPs divide the dataset into training and calibration sets during the model-building process. The training set is used to develop the model, while the calibration set helps calculate the nonconformity scores. This two-step process optimizes computation and delivers precise prediction regions.

Conformal Prediction for Calibrated Uncertainty

A set prediction that guarantees provable coverage of the ground truth with a user-specified probability

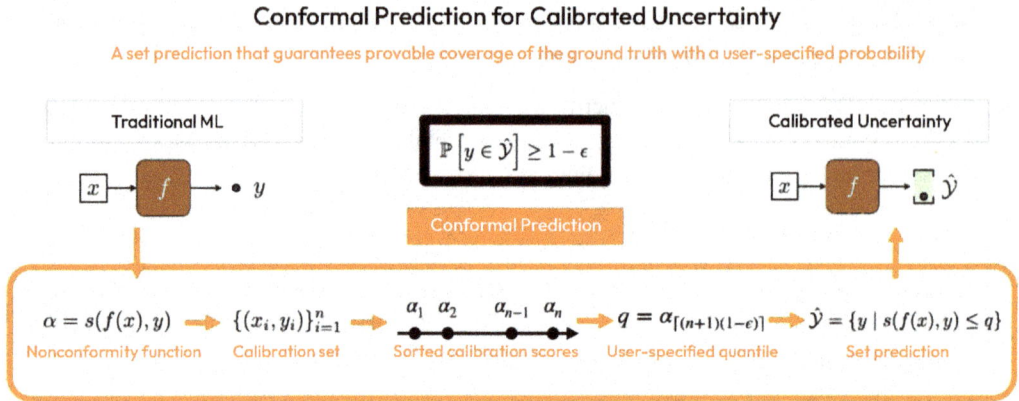

Figure 5.3 – Inductive conformal prediction

A predictive model, such as a neural network or a decision tree, is first trained on the proper training set. Then, the nonconformity of each example in the calibration set is computed using the trained model. The nonconformity measure is a real-valued function that describes how much an example contradicts the rest of the data. The nonconformity scores of the calibration set are then used to determine the size of the prediction region for new examples.

The inductive approach offers a significant computational advantage, particularly for large datasets. By creating the predictive model only once, ICP reduces the algorithm's time complexity, unlike TCP, which requires retraining the model for each new prediction. However, it's important to note that ICP assumes the data are exchangeable, meaning the data's order doesn't carry any information.

In terms of applications, inductive conformal predictors can be used for both classification (binary and multiclass) and regression tasks. The method offers a flexible and efficient way of providing a measure of uncertainty associated with predictions, a valuable feature in many practical applications.

ICP involves several steps, most of which center around the calculation of nonconformity scores. Here's a rough outline of the algorithm along with the associated mathematical formulation:

1. **Data partitioning**: Split the initial dataset D into a proper training set D_train, and a calibration set D_cal.

2. **Model training**: Train a predictive model M on D_train. This model is used to generate predictions on new instances. The type of model (e.g., SVM, decision tree, linear regression, etc.) depends on the problem at hand.

3. **Nonconformity measure calculation**: Use the trained model M to predict outcomes for instances in the calibration set D_cal. For each instance of, (x_i, y_i) in D_cal, compute a nonconformity score α_i, representing the *strangeness* or *abnormality* of the instance. The nonconformity measure α is generally problem-specific. For instance, classification tasks could be hinge loss $1 - p\_yi$, where p_yi is the predicted probability of the correct class y_i according to the model M. For regression, it could be the absolute error $|y\_i - y\_hat\_i|$, where y_hat_i is the model's prediction for x_i.

4. **Prediction interval generation**: Given a new test point x, calculate its nonconformity score α_x using model M. Then, calculate the p-value for x, which is the proportion of instances in D_cal with nonconformity scores greater than or equal to α_x. p_x can be computed as follows:

$$p\,x = (1 + |i{:}\alpha\,i \geq \alpha\,x|)/(|Dcal| + 1)$$

Here, $|\{i: \alpha\_i \geq \alpha\_x\}|$ denotes the number of instances in the calibration set with nonconformity scores greater than or equal to α_x. This p-value represents how often we expect to observe a nonconformity score at least as large as α_x by chance.

5. **Prediction output**: Using the calculated p-value, create a prediction set $\Gamma(x)$ for the new test point x. For classification, the prediction set contains all classes y for which the p-value is at least the chosen significance level ε: $\Gamma(x) = \{y: p\_y \geq \varepsilon\}$. For regression, an interval prediction (y_lower, y_upper) is typically outputted, where y_lower and y_upper are the lowest and highest values, respectively, for which the p-value is at least as large as the chosen significant level.

Please note that this is a high-level description of the algorithm and mathematical formulation. The exact details may vary based on the specific form of ICP used and the type of problem (classification, regression, etc.) being addressed.

As we've unpacked the complexities and capabilities of both classical and inductive approaches, it's now essential to discern how to choose the optimal method for a given situation. Let's navigate the factors and guidelines that will guide you in selecting the best-fit conformal predictor for your specific needs in the upcoming section, *Choosing the right conformal predictor*.

Choosing the right conformal predictor

Both classical and inductive conformal predictors offer valuable approaches to building reliable machine learning models. However, they each come with unique strengths and weaknesses.

Classical transductive conformal predictors are highly adaptable and do not make any assumptions about data distribution. However, they tend to be computationally expensive, requiring the model's retraining for each new prediction.

Inductive conformal predictors, conversely, are computationally more efficient, as they only require the model to be trained once.

Choosing the right conformal predictor largely depends on the specific requirements of the problem at hand. Some considerations might include the following:

- **Computation resources**: If computation resources or time are a concern, inductive conformal predictors might be more suitable due to their reduced computational cost

- **Data size**: For smaller datasets, classical conformal predictors might be more suitable, while for larger datasets, inductive conformal predictors are usually the preferred choice due to computational efficiency

- **Data quality**: If data quality is high, inductive conformal predictors can be a good choice

- **Real-time requirements**: If the model needs to make real-time predictions, inductive conformal predictors might be more suitable due to their one-time training process

Here are real-life scenarios illustrating when one might opt for transductive or inductive conformal predictors.

Transductive conformal predictors

medical diagnostics with limited data:

- **Scenario**: A hospital uses machine learning to diagnose a rare disease but only has a limited dataset of past patients.

- **Reasoning**: Given the smaller dataset and the critical nature of accurate predictions, classical TCP is favored. Its adaptability and distribution-free nature might lead to more accurate predictions, even if it requires more computational power per prediction.

Inductive conformal predictors

e-commerce recommendation systems:

- **Scenario**: A large e-commerce platform wants to provide real-time product recommendations to millions of its users based on their browsing habits.

- **Reasoning**: Due to the massive scale, the system can't afford to retrain models for every recommendation. ICP's one-time training process, combined with its computational efficiency for larger datasets, makes it a suitable choice.

To effectively choose the appropriate type of conformal predictor, it's essential to gain a deep understanding of both classical and inductive conformal predictors, their working principles, and their strengths and weaknesses. Furthermore, understanding the nature and requirements of the problem domain, such as the specific characteristics of the data, the computational resources available, the need for real-time predictions, and the importance of model interpretability, can significantly aid in making an informed choice. Always remember that the best conformal predictor is the one that best meets the needs of your specific problem domain.

Summary

This chapter explored the fascinating world of conformal predictors, their types, and their distinctive features. The key concepts and skills we touched upon include covering the foundational principles of conformal prediction and its application in machine learning. It also highlighted the differences between classical transductive and inductive conformal predictors. We also covered how to effectively choose the appropriate type of conformal predictor based on the specific requirements of the problem. Finally, the practical applications of conformal predictors in binary classification, multiclass classification, and regression were also included.

The chapter also provided a detailed algorithmic description and mathematical formulation of classical and inductive conformal predictors, adding to our theoretical understanding. To deepen our learning, we also took a hands-on approach, looking at practical examples in Python.

For those interested in further exploring conformal predictors, several avenues exist for you to consider. A more detailed study of the mathematical underpinnings of conformal prediction could be pursued, along with implementing conformal predictors in more complex machine learning models.

Exploring the advanced versions of conformal predictors, such as Mondrian conformal predictors, or understanding how conformal prediction can be integrated with other machine learning techniques, such as neural networks and ensemble learning, are also exciting areas for further research.

In closing, we hope this chapter has given a solid grounding in the principles and applications of conformal prediction. Moving into the next chapter, we'll delve deeper into conformal prediction for classification problems. As always, keep exploring, keep learning, and enjoy the journey!

Part 3: Applications of Conformal Prediction

In this part, we will provide more details about conformal prediction for classification problems. It will introduce the calibration concept and illustrate how conformal prediction compares with other calibration methods, explaining how it can quantify uncertainty in regression to produce well-calibrated prediction intervals. This part will also explain how conformal prediction can produce prediction intervals for point forecasting models, illustrate applications using open source libraries, and detail recent innovations in conformal prediction for NLP. Finally, this part will explain how conformal prediction can be applied to produce state-of-the-art uncertainty quantification for NLP and illustrate applications using open source libraries.

This section has the following chapters:

- *Chapter 6, Conformal Prediction for Classification*
- *Chapter 7, Conformal Prediction for Regression*
- *Chapter 8, Conformal Prediction for Time Series and Forecasting*
- *Chapter 9, Conformal Prediction for Computer Vision*
- *Chapter 10, Conformal Prediction for Natural Language Processing*

6

Conformal Prediction for Classification

This chapter dives deeper into the topic of conformal prediction for classification problems. We will explore the concept of classifier calibration and demonstrate how conformal prediction compares to other calibration methods before introducing Venn-ABERS predictors as specialized techniques within conformal prediction. Additionally, we will provide an overview of open source tools that can be utilized to implement conformal prediction for classifier calibration.

We will cover the following topics in this chapter:

- Classifier calibration
- Evaluating calibration performance
- Various approaches to classifier calibration
- Conformal prediction for classifier calibration
- Open source tools for conformal prediction in classification problems

Classifier calibration

Most statistical, machine learning, and deep learning models output predicted class labels, and the models are typically evaluated in terms of their accuracy.

Accuracy is a prevalent measure for assessing the performance of a machine learning classification model. It quantifies the ratio of instances that are correctly identified to the overall count in the dataset. In other words, accuracy tells us how often the model's predictions align with the true labels of the data.

The accuracy score measures how often the model's predictions match the true observed labels. It is calculated as the fraction of correct predictions out of all predictions made. Accuracy scores between 0 and 1 quantify how accurate the model's predictions are compared to the ground truth data. A higher accuracy score close to 1 signifies that the model is performing very accurately overall, with most of its predictions being correct. A lower accuracy approaching 0 indicates poor performance, with the majority of the model's predictions being incorrect compared to the true labels.

The closer the accuracy is to 1, the better the model is performing. The closer it is to 0, the worse the model is at predicting the true labels in the data.

Accuracy is a straightforward and intuitive metric that is easy to understand and interpret. However, it may not always be the most suitable metric, especially when dealing with imbalanced datasets. In imbalanced datasets, where the number of instances in different classes is significantly different, accuracy alone may be misleading. In an imbalanced dataset, a classifier that consistently predicts the majority class can attain a high accuracy based on the class distribution, even if it doesn't identify the minority class.

In these scenarios, it's crucial to look at other evaluation measures, such as precision, recall, F1 score, or ROC-AUC, to gain a fuller insight into the model's efficacy.

Depending on the specific problem and the requirements, other metrics and considerations, such as the cost of false positives or false negatives, might be more relevant. Therefore, it is essential to assess the model's performance using multiple evaluation metrics and consider the context in which the classification model will be applied.

Accuracy alone may be insufficient, particularly in critical applications, for several reasons:

- **Imbalanced datasets**: In scenarios where the dataset is imbalanced, accuracy can be misleading. If the majority class dominates the dataset, a model that predicts only the majority class can achieve high accuracy but fails to capture the minority class effectively. This can be problematic in critical applications where correctly identifying rare events or detecting anomalies is crucial.

- **Cost of errors**: In many real-world applications, the cost of false positives and false negatives can vary significantly. Accuracy treats all errors equally and does not consider the consequences of misclassifications. For instance, in a medical diagnosis, a false negative (failing to detect a disease) can be far more critical than a false positive. In such cases, accuracy alone does not provide sufficient information about the model's performance in terms of the actual impact on decision-making and outcomes.

- **Probability estimation**: Accuracy does not take into account the confidence or uncertainty of the model's predictions. It is essential to assess the model's ability to provide well-calibrated probability estimates. Calibration refers to the alignment between predicted probabilities and the true probabilities of events. A poorly calibrated model may provide overly confident or unreliable probability estimates, which can lead to incorrect decisions or the misinterpretation of risks.

- **Decision threshold**: Accuracy does not consider the decision threshold used for classification. Different decision thresholds can result in varying trade-offs between precision and recall. Depending on the application, certain misclassification errors may be more tolerable than others. Evaluating only accuracy does not provide insights into the model's performance at different decision thresholds.

Let's get to understanding the concepts of classifier calibration next.

Understanding the concepts of classifier calibration

In the previous chapters, we defined and discussed the concept of classifier calibration.

Classifier calibration involves adjusting the predicted probabilities from a classification model so that they better reflect the true likelihood of each class. The goal is to make the predictions better calibrated.

A well-calibrated classifier is one where the predicted probabilities match the empirical probabilities. For example, if the model predicts "class A" with 60% probability across 100 examples, then class A should occur approximately 60 times out of those 100 predictions.

More formally, a well calibrated classifier satisfies the following formula:

$$P(actual\ class\ is\ c \mid predicted\ probability\ of\ c\ is\ p) \approx p$$

This means the observed frequency of class c should be close to p when the model predicts class c with probability p.

Calibration adjustment ensures the predicted probabilities are aligned with the relative frequencies in the actual data. The predictions are calibrated to the empirical evidence so that a predicted probability of 0.7 corresponds to a 70% chance based on the data. This calibration is essential for probability estimates to be meaningful and reliable.

For instance, consider a binary classifier that predicts whether an email is spam or not. For each email, it might predict a probability, say, 0.8, which means it believes there's an 80% chance that the email is spam. If the classifier is well calibrated, then out of all emails that it assigns a spam probability of 0.8, about 80% should be spam.

Without calibration, the output probabilities of a classifier might not correspond to the true likelihood of the predicted class, which can be problematic for decision-making. Calibration methods adjust these probabilities to better reflect reality. The goal is to have the output probabilities of the classifier be as close as possible to the true probabilities.

Model calibration is crucial because of the following aspects:

- **Reliable probability estimates**: Calibrated classifiers provide accurate and reliable probability estimates for the predicted classes. Probability estimates reflect the model's confidence in its predictions and can be interpreted as the likelihood of a particular class being correct. In many real-world applications, such as medical diagnosis, risk assessment, or fraud detection, having well-calibrated probability estimates is crucial for making informed decisions and assessing the level of uncertainty associated with the predictions.

- **Reliable risk assessment**: In many domains, accurate risk assessment is paramount. Calibrated classifiers provide well-calibrated probability estimates that reflect the true likelihood of events. This allows for more accurate and reliable risk assessment, enabling decision-makers to allocate resources, prioritize actions, or estimate the impact of certain events more effectively. For instance, in credit scoring, a calibrated classifier can provide accurate estimates of the probability of default, aiding in better risk management.

- **Decision threshold determination**: In classification tasks, decisions are often made by setting a threshold on the predicted probabilities. This threshold determines the trade-off between precision and recall, or equivalently, between false positives and false negatives. Calibrated classifiers help in selecting an appropriate decision threshold by aligning the probability estimates with the desired trade-off, considering the specific costs or consequences associated with different types of errors. This ensures that decision-making aligns with the objectives and requirements of the application.

- **Interpretability and trust**: Calibration enhances the interpretability of the model's predictions. Calibrated probability estimates can be used to understand the level of confidence the model has in its predictions. This transparency helps in building trust with users, stakeholders, and regulatory authorities, particularly in domains where decision-making is critical and must be justified. By providing well-calibrated probability estimates, the model's predictions can be better understood and validated, instilling confidence in its reliability.

- **Improved fairness**: Calibrated classifiers can contribute to fairness in decision-making processes. By providing well-calibrated probability estimates, they can help in identifying and mitigating biases that may arise from the underlying training data or model assumptions. This allows for fairer and more equitable predictions, ensuring that different groups are treated consistently and without undue bias.

It is essential to evaluate model calibration to ensure that the model's predictions align with the underlying uncertainties in the data. This evaluation helps in making informed decisions, understanding the model's limitations, and managing the risks associated with misclassifications or incorrect probability estimates.

Traditionally, many classifiers, such as logistic regression or support vector machines, generate probability estimates based on their internal models. However, these probability estimates are not always accurate or well calibrated, leading to overconfidence or under confidence in the predictions. For instance, a classifier may assign probabilities close to 1.0 or 0.0 to certain examples when it should have assigned probabilities closer to 0.7 or 0.3, respectively.

To address this issue, various techniques have been proposed to calibrate classifiers and improve the reliability of their probability estimates. These techniques aim to map the original probability scores to more accurate and calibrated probabilities. The goal is to ensure that, on average, the predicted probabilities match the observed frequencies or likelihoods of the predicted events.

Evaluating calibration performance

Evaluating the calibration performance of a classifier is crucial to assessing the reliability and accuracy of its probability estimates. Calibration evaluation allows us to determine how well the predicted probabilities align with the true probabilities or likelihoods of the predicted events. Here are some commonly used techniques for evaluating the calibration performance of classifiers:

- **Calibration plot**: A calibration plot visually assesses how well a classifier's predicted probabilities match the true class frequencies. The x axis shows the predicted probabilities for each class, while the y axis shows the empirically observed frequencies for those predictions.

  For a well-calibrated model, the calibration curve should closely match the diagonal, representing a 1:1 relationship between predicted and actual probabilities. Deviations from the diagonal indicate miscalibration, where the predictions are inconsistent with empirical evidence.

  Calibration plots provide an intuitive way to identify if a classifier is over confident or under confident in its estimates across different probability ranges. The closer the curve aligns with the diagonal, the better calibrated the predicted probabilities are. Significant deviations signal that recalibration is needed for the model's outputs to be reliable:

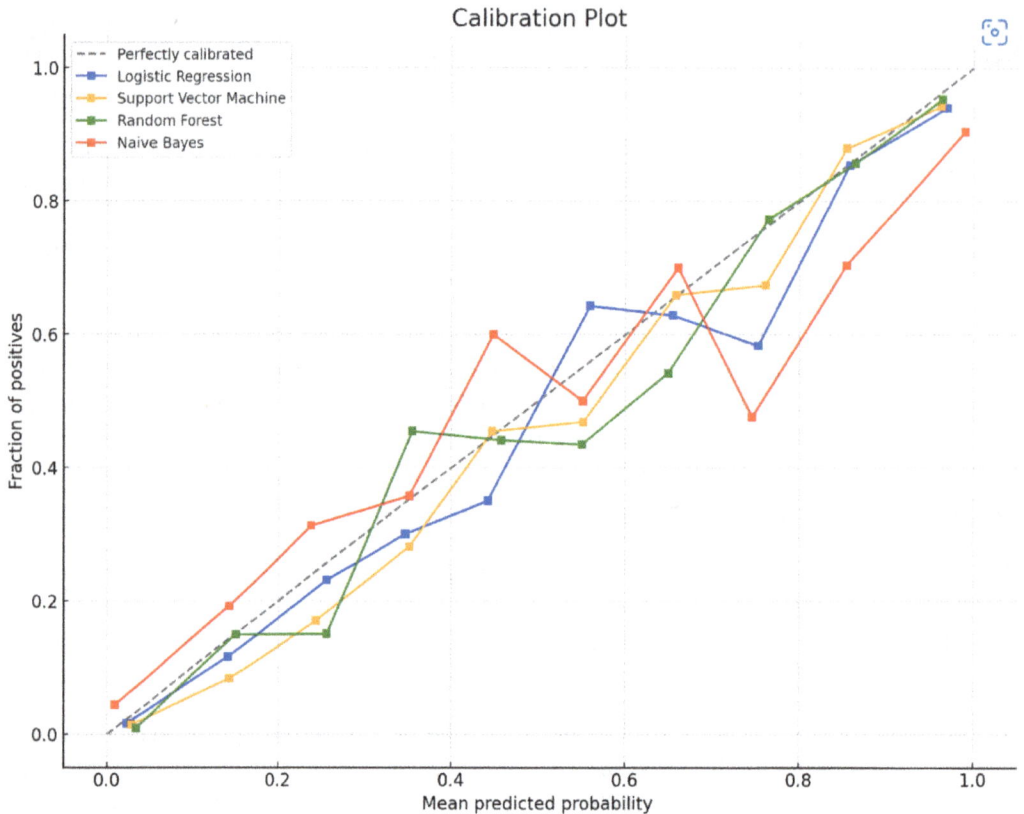

Figure 6.1 – Calibration plot

- **Calibration error**: Calibration error measures the average difference between the predicted probabilities and the actual probabilities of the forecasted events. It's determined by the mean absolute deviation between the estimated probabilities and the observed probabilities. Lower calibration error values indicate better calibration performance.

- **Calibration metrics**: Several metrics can be used to evaluate the calibration performance of a classifier. Commonly used metrics include the **expected calibration error** (ECE), log loss, and the Brier score. ECE measures the calibration error by partitioning the predicted probabilities into bins and calculating the difference between the average predicted probabilities and the average empirical probabilities within each bin. The Brier score assesses the overall accuracy of the predicted probabilities, considering both calibration and resolution (sharpness) of the probability estimates. The Brier score is a commonly used scoring rule for assessing the calibration of probabilistic forecasts. It measures the mean squared difference between the predicted probabilities and the actual outcomes.

For a set of N predictions, the Brier score is calculated as $BS = \frac{1}{N}\sum_{t=1}^{N}(f_t - o_t)^2$, where $f_{t,i}$ is the forecasted probability for an event, i, at time t, and $o_{t,i}$ is the actual outcome of an event, i, at time t (0 or 1).

Squaring the errors gives more weight to large mistakes. The average squared error is then taken across all predictions.

A lower Brier score indicates better calibration, with a minimum of 0 for a perfect probabilistic forecaster. It penalizes both inaccurate and over/underconfident predictions.

- **Cross-validation**: Cross-validation is a technique for estimating the calibration performance. It does this by partitioning the dataset into multiple folds and training the model on one fold while evaluating calibration on the remaining folds. This helps in assessing the calibration performance across different subsets of the data and provides a more robust evaluation.

When evaluating the calibration performance, it is important to compare the results against an appropriate baseline. A well-calibrated classifier should outperform random or uncalibrated probability estimates.

Various approaches to classifier calibration

Before exploring how conformal prediction can provide calibrated probabilities, we will first discuss some common non-conformal calibration techniques and their strengths and weaknesses. These include histogram binning, Platt scaling, and isotonic regression.

It is important to note that the following methods are not part of the conformal prediction framework. We are covering them to build intuition about calibration and highlight some of the challenges with conventional calibration approaches. This background will motivate the need for and benefits of the conformal prediction perspective so that we can obtain reliable probability estimates.

The calibration techniques we will explore, including histogram binning, Platt scaling, and isotonic regression, represent widely used approaches for adjusting classifier confidence values. However, as we will discuss, they have certain limitations regarding model flexibility, computational expense, and generalization.

By first understanding these existing calibration methods and their drawbacks, we will be equipped to better comprehend the value of conformal prediction's inherent calibration properties. This background provides context into the calibration problem before presenting conformal prediction as an attractive modern solution.

Histogram binning

Histogram binning is a technique that's commonly used in classifier calibration to improve the calibration performance of probability estimates. It involves dividing the predicted probabilities into bins or intervals and mapping them to more accurate and reliable probabilities based on the empirical frequencies or observed proportions of the predicted events within each bin. The goal of histogram binning is to align the predicted probabilities with the true probabilities of the events, resulting in a better-calibrated classifier.

The process of histogram binning can be summarized as follows:

1. **Partitioning**: The predicted probabilities are divided into a predefined number of bins or intervals. The number of bins can vary based on the dataset and the desired granularity of calibration.

2. **Bin assignment**: Each instance in the dataset is assigned to the corresponding bin based on its predicted probability. For example, if we have five bins with equal width intervals (for example, *0-0.2*, *0.2-0.4*, *0.4-0.6*, *0.6-0.8*, and *0.8-1.0*), an instance with a predicted probability of 0.45 would be assigned to the third bin.

3. **Calibration mapping**: Within each bin, the empirical proportion or frequency of the true events (positives) is calculated. This can be obtained by computing the ratio of the number of positive instances to the total number of instances within the bin. For instance, if the third bin contains 100 instances, and 70 of them are true positives, the empirical proportion of positives within that bin would be 0.7.

4. **Mapping to calibrated probabilities**: The predicted probabilities within each bin are then mapped or adjusted to more accurate and calibrated probabilities based on the empirical proportions of positives. This mapping can be performed using various techniques, such as isotonic regression or Platt scaling.

5. **Overall calibration**: Once the mapping has been applied to all the bins, the calibrated probabilities are obtained by combining the probabilities from all the bins. The result is a set of calibrated probabilities that better align with the true probabilities or likelihoods of the events.

Here are some potential disadvantages of using histogram binning for classifier calibration:

- **Inflexibility**: Histogram binning divides the prediction space into fixed intervals. It lacks the flexibility to model complex, nonlinear miscalibration patterns.

- **Data underutilization**: Hard binning discards information within each bin. The calibration mapping uses only the bin averages rather than the full distribution.

- **Sensitivity to the binning scheme**: The calibration quality is dependent on the specific binning thresholds chosen, which can be arbitrary. Optimal binning is often not known beforehand.

- **Discontinuities**: Adjacent bins may have very different adjustments, leading to abrupt discontinuities in the calibration mapping. This can introduce artifacts.

- **Difficulty extrapolating**: The binning calibration is based only on the training data distribution. It may not extrapolate well to unseen data with sparse or no coverage.

- **Curse of dimensionality**: Histograms do not scale well to high-dimensional feature spaces. The data becomes too sparse within each bin.

- **Limited model expressiveness**: Histograms can only represent simple, low-order calibration relationships. They cannot model complex miscalibration patterns.

Histogram binning can be simple to implement but provides an inflexible, discontinuous calibration mapping. More sophisticated dense modeling and smoothing are often required for optimal calibration quality.

Platt scaling

Platt scaling, sometimes referred to as Platt's method or sigmoid calibration, is a post-processing approach that's employed to refine the output probabilities of a binary classification model. It was introduced by John C. Platt in 1999 to transform the raw output scores of a support vector machines classifier into well-calibrated probabilities.

The goal of Platt scaling is to adjust the predicted scores or logits produced by the classifier in such a way that they reflect more accurate estimates of the true probabilities. This is achieved by fitting a logistic regression model on the classifier's scores while using a labeled validation set or a holdout set. The logistic regression model is trained to map the original scores to calibrated probabilities.

The steps involved in Platt scaling are as follows:

1. Collect a labeled validation set or a holdout set that is distinct from the training data used to train the classifier.

2. Use the classifier to generate the raw output scores or logits for the instances in the validation set.

3. Fit a logistic regression model on the validation set, treating the raw scores as the independent variable and the true class labels as the dependent variable.

4. Train the logistic regression model using standard techniques such as maximum likelihood estimation or gradient descent to estimate the model's parameters.

5. Once the logistic regression model has been trained, it can be used as a calibration function. Given a new instance, the raw score produced by the classifier is input into the logistic regression model, which transforms it into a calibrated probability estimate.

The logistic regression model essentially learns the transformation from the raw scores to calibrated probabilities by estimating the intercept and slope parameters. This transformation is represented by the sigmoid function, which maps the scores to probabilities between 0 and 1.

Platt scaling aims to achieve better calibration by adjusting the predicted probabilities to match the true probabilities or likelihoods of the events.

It's important to note that Platt scaling assumes that the relationship between the raw scores and the true probabilities can be modeled by a logistic function. If the underlying relationship is more complex, other calibration methods such as conformal prediction may be more suitable.

While Platt scaling can be an effective technique for calibrating classifier probabilities, it is important to be aware of its limitations and potential disadvantages:

- **Requirement for a separate validation set**: Platt scaling requires a labeled validation set or holdout set that is distinct from the training data. This means additional data may be needed for calibration, which can be a limitation in situations where obtaining labeled data is challenging or costly.

- **Assumption of a logistic relationship**: Platt scaling assumes that the relationship between the raw scores and the true probabilities can be accurately modeled by a logistic function. If the underlying relationship is more complex or different, the logistic regression model may not be able to capture the true calibration mapping adequately.

- **Sensitivity to extreme scores**: Platt scaling can be sensitive to extreme scores or outliers in the validation set. Outliers may disproportionately influence the calibration function, leading to potential overfitting or a suboptimal calibration.

- **Lack of flexibility for different calibration shapes**: The logistic regression model used in Platt scaling is constrained to fit a sigmoid function, which may not be suitable for all calibration shapes. If the desired calibration shape deviates significantly from a sigmoid curve, Platt scaling may not achieve optimal calibration.

- **Limited applicability to multiclass problems**: Platt scaling is primarily designed for binary classification problems. Extending it to multiclass classification can be challenging as it requires adapting the calibration mapping to handle multiple classes and their respective probabilities.

- **Potential overconfidence in extreme probabilities**: Platt scaling may introduce overconfidence in extreme predicted probabilities. The calibrated probabilities near the boundaries (close to 0 or 1) might be more extreme than they should be, leading to overconfident predictions in those regions.

- **Dependence on the quality of the validation set**: The effectiveness of Platt scaling is dependent on the quality and representativeness of the labeled validation set. If the validation set does not accurately capture the true distribution of the target variable, the resulting calibration may be suboptimal.

It's important to consider these disadvantages and assess whether Platt scaling is suitable for a specific application or if alternative calibration methods, such as methods based on conformal prediction, may be more appropriate based on the specific characteristics of the problem and dataset.

Isotonic regression

Isotonic regression is a non-parametric regression technique that's used for calibration and monotonicity modeling. It is commonly applied to adjust the output scores or predicted probabilities of a classifier to improve their calibration. Isotonic regression seeks to find a monotonic function that maps the original scores to calibrated probabilities while preserving the ordering of the scores.

The primary goal of isotonic regression is to determine a non-decreasing function that reduces the sum of squared discrepancies between the predicted probabilities and the target probabilities or actual occurrences. By fitting a piecewise linear or piecewise constant function to the data, isotonic regression ensures that the predicted probabilities are monotonically increasing or non-decreasing.

The steps involved in isotonic regression are as follows:

1.   Collect a labeled validation set or a holdout set that is separate from the training data.

2.   Use the classifier to generate the raw output scores or probabilities for the instances in the validation set.

3.   Sort the instances in the validation set based on the raw scores.

4.   Initialize the isotonic regression function as the identity function, where the initial predicted probabilities are equal to the raw scores.

5.   Iteratively update the isotonic regression function by adjusting the predicted probabilities to minimize the squared differences between the predicted probabilities and the target probabilities. This adjustment is subject to the constraint of non-decreasing probabilities.

6.   Repeat the updating process until convergence or a stopping criterion is reached:

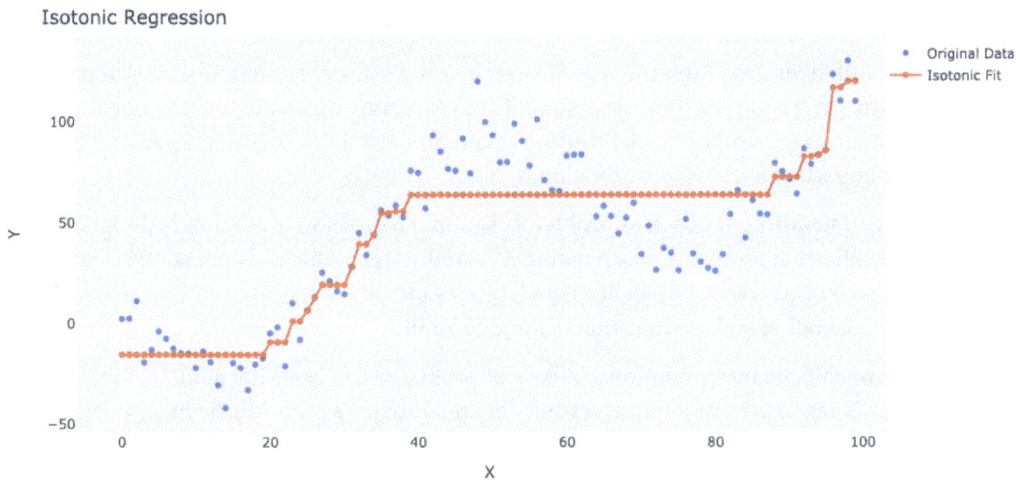

Figure 6.2 – Isotonic regression

Once the isotonic regression model has been trained, it can be used to map the raw scores of new instances to calibrated probabilities. The model ensures that the predicted probabilities are monotonically increasing and better aligned with the true probabilities or likelihoods of the events.

While isotonic regression is a valuable technique for calibrating classifier probabilities, it is important to consider its limitations and potential disadvantages:

- **Potential overfitting**: Isotonic regression can suffer from overfitting if the calibration function is overly complex or if the calibration dataset is small. Regularization techniques, such as using a limited number of segments in the piecewise linear function, can help prevent overfitting.

- **Complexity and computational cost**: Isotonic regression can be resource-intensive, especially with vast datasets or when navigating high-dimensional features. As the number of data points and unique scores or probabilities grow, so does the complexity of isotonic regression. It's crucial to weigh up the computational limitations when using isotonic regression for extensive tasks.

- **Sensitivity to outliers**: Isotonic regression can be sensitive to outliers in the data. Outliers may have a significant impact on the estimated calibration function, potentially leading to suboptimal calibration. Careful data preprocessing or outlier detection techniques may be necessary to mitigate this issue.

- **Limited flexibility for complex calibration shapes**: Isotonic regression assumes a monotonic relationship between the scores and the probabilities, which constrains the calibration function to be piecewise constant or piecewise linear. This limits the model's flexibility to capture more complex or nonlinear calibration shapes. If the desired calibration shape deviates significantly from monotonicity, isotonic regression may not provide an optimal fit.

- **Need for sufficient data**: Isotonic regression requires a sufficient amount of labeled data to estimate the calibration function accurately. If the calibration dataset is small or imbalanced, the estimation may be suboptimal. Ensuring a representative and adequately sized calibration dataset is important for reliable calibration results.

- **Difficulty in handling multiclass problems**: Isotonic regression is inherently designed for binary classification problems, so extending it to multiclass problems is not straightforward. Adapting isotonic regression to handle multiple classes and their respective probabilities requires careful consideration and modification of the algorithm.

- **Lack of probabilistic interpretation**: Unlike Platt scaling, which explicitly models probabilities using logistic regression, isotonic regression does not provide a probabilistic interpretation of the calibrated scores. It focuses solely on ensuring monotonicity and may not directly estimate well-calibrated probabilities.

It is important to evaluate these limitations and consider the characteristics of the problem at hand when deciding whether isotonic regression is the most appropriate calibration method.

Comparing the advantages and disadvantages of different calibration techniques, such as conformal prediction and Platt scaling, can help determine the best approach for achieving well-calibrated probabilities in a specific application.

Conformal prediction for classifier calibration

Conformal prediction is a powerful framework for probabilistic prediction that provides valid and well-calibrated prediction sets and prediction intervals. It offers a principled approach to quantify and control the uncertainty associated with the predictions.

We have already seen how conformal prediction approaches, such as **inductive conformal prediction** (**ICP**) and **transductive conformal prediction** (**TCP**), aim to generate sets that have accurate coverage probabilities. To recap, conformal prediction computes p-values and constructs prediction sets by comparing the p-values of each potential label with a selected significance level.

Unlike Platt scaling, histogram binning, and isotonic regression, which focus on calibrating the predicted probabilities or scores, conformal prediction takes a more comprehensive approach by providing prediction sets that encompass the uncertainty associated with the predictions and enhances the reliability and interpretability of predictions by providing valid measures of confidence or significance.

Venn-ABERS conformal prediction

Classical methods such as Platt scaling was initially developed as parametric solutions for calibrating classifiers. However, these methods are becoming somewhat outdated and have limitations due to their simplistic assumptions, resulting in suboptimal calibration of probabilities.

Platt scaling assumes a logistic relationship between scores and probabilities, which may not adequately capture the actual calibration shape in practical scenarios. It is worth noting that Platt's original paper in 1999 did not explicitly discuss the underlying assumptions of this approach. However, recent research (see *Beta calibration: a well-founded and easily implemented improvement on logistic calibration for binary classifiers*: `https://proceedings.mlr.press/v54/kull17a.html`) has revealed that these assumptions are essentially equivalent to assuming both normality and homoscedasticity, which are overly restrictive assumptions for real-world datasets. Real datasets often exhibit more complex and diverse patterns that cannot be accurately captured by such assumptions. Therefore, relying solely on Platt scaling with its underlying assumptions may result in suboptimal calibration and poorly calibrated probabilities.

Isotonic regression, as an approach, assumes a monotonic relationship between scores and probabilities. However, this assumption may not capture the intricate nature of the calibration curve in all cases. Furthermore, isotonic regression relies on the assumption of perfect ranking (an ROC AUC of 1) on the test dataset, which is rarely achievable in real-world datasets. Additionally, it has been demonstrated that isotonic regression can overfit when applied to smaller datasets.

The assumption of a monotonic relationship limits the flexibility of isotonic regression to model more complex calibration curves that may exhibit non-monotonic patterns. Moreover, the requirement of perfect ranking on the test dataset is often unrealistic as datasets typically involve inherent noise and uncertainty. This assumption can lead to suboptimal calibration results in practice.

Furthermore, the issue of overfitting with isotonic regression becomes more prominent when dealing with smaller datasets. When the dataset's size is limited, isotonic regression may overly adjust to the noise or specific characteristics of the training data, resulting in poor generalization performance.

Because of their simplistic assumptions, Platt scaling and isotonic regression may not achieve optimal calibration and may not deliver well-calibrated probabilities. These methods may struggle to capture nonlinear or more intricate calibration patterns, limiting their effectiveness in certain applications.

To address the limitations of classical calibrators, such as Platt scaling and isotonic regression, a powerful solution called Venn-ABERS has been developed by the creator of conformal prediction, Vladimir Vovk , Ivan Petej and Valentina Fedorova", Venn-ABERS is a conformal prediction method that offers mathematical guarantees of validity, regardless of the data distribution, dataset size, or underlying classification model.

This work is detailed in the NeurIPS paper titled *Large-scale probabilistic predictors with and without guarantees of validity* (`https://papers.nips.cc/paper_files/paper/2015/hash/a9a1d5317a33ae8cef33961c34144f84-Abstract.html`). For a more mathematical understanding, watch the associated presentation, *Large-Scale Probabilistic Prediction With and Without Validity Guarantees*, at `https://www.youtube.com/watch?v=ksrUJdb2tA8`.

The name Venn-ABERS is derived from a combination of Venn predictors, another class of conformal predictors, and the initials of the authors who contributed to a classical paper called *An Empirical Distribution Function for Sampling with Incomplete Information* (M. Ayer, H.D. Brunk, G.M. Ewing, W.T. Reid, and E. Silverman: `https://projecteuclid.org/journals/annals-of-mathematical-statistics/volume-26/issue-4/An-Empirical-Distribution-Function-for-Sampling-with-Incomplete-Information/10.1214/aoms/1177728423.full`).

So, how do Venn-ABERS predictors work? Rather than constructing isotonic regression once, Venn-ABERS fits isotonic regression twice by assuming that each test object can have both label 0 and label 1. This means that each test object is added to the calibration set twice, once with label 0 and once with label 1. Two separate isotonic regressions are then fitted, resulting in two probabilities, *p0* and *p1*, for each test object.

It is important to note that both *p0* (lower bound) and *p1* (upper bound) represent probabilities of the object belonging to class 1. These probabilities create a prediction interval for the probability of class 1, with mathematical guarantees that the actual probability falls within this interval.

Consequently, Venn-ABERS solves the problem beautifully, without requiring assumptions about score distributions such as Platt scaling and without suffering from overfitting.

The Venn-ABERS prediction is a multi-predictor, and the width of the interval (*p0, p1*) contains valuable information about the confidence of classification. In larger datasets, *p0* and *p1* are typically very close to each other. However, for smaller and more challenging datasets, *p0* and *p1* may diverge, indicating that certain objects are difficult to classify due to factors such as data distribution, insufficient data, or the underlying classifier's performance.

Importantly, in critical situations, the Venn-ABERS predictor not only outputs accurate and well-calibrated probabilities but also issues an "alert" by widening the $(p0, p1)$ interval. This alert indicates that the decision-making process should consider the increased uncertainty.

For practical decision-making purposes, the probabilities can be combined into a single value using $p = p1 / (1 - p0 + p1)$. This combined probability of class 1, p, can then be utilized for decision-making tasks such as loan granting or determining whether to disable autopilot in autonomous car. With the inclusion of p, the decision-making process can be successfully concluded.

Comparing calibration methods

Given the range of calibration methods, you might be wondering how they are compared to each other. We have already seen that classical methods such as Platt scaling and isotonic regression rely on restrictive assumptions and, unlike the conformal prediction Venn-ABERS method, do not have validity guarantees. An interesting question is also how the performance of different methods compares empirically across a range of datasets.

Such a study was performed, and the results were summarized in the paper *Probabilistic Prediction in scikit-learn* (http://www.diva-portal.org/smash/get/diva2:1603345/FULLTEXT01. pdf). In this paper, a large experimental study was conducted to investigate the calibration of scikit-learn models out of the box. In addition, the study looked at whether calibration techniques such as Platt scaling, isotonic regression, and Venn-ABERs can improve calibration.

The result of the study showed that of the seven algorithms evaluated (logistic regression, random forest, AdaBoost, gradient boosting, kNN, naïve Bayes, and decision tree), the only model that obtained well-calibrated predictions was logistic regression. Calibration enhances all models, especially decision trees, boosted trees (such as XGBoost, LightGBM, and CatBoost), and naïve Bayes. This underscores the clear advice for professionals: obtained relatively well-calibrated predictions was logistic regression.

Additionally, the study uncovered a notable finding that miscalibrated models tend to exhibit a high level of overconfidence. Surprisingly, even logistic regression, although to a lesser extent compared to other models, displayed systematic optimism in its predictions.

In other words, these miscalibrated models tended to assign higher probabilities or confidence to their predictions than what was warranted by the actual outcomes. This overconfidence could potentially lead to misguided decision-making or misplaced trust in the reliability of the predictions.

It is crucial to recognize that while logistic regression demonstrated better calibration compared to other models, it still exhibited a certain level of systematic optimism. This highlights the importance of thoroughly evaluating and calibrating models, even those considered to be well-calibrated, to ensure accurate and reliable probabilistic predictions.

When examining the calibration techniques in terms of their benefits for calibration, their order of effectiveness is typically observed to be Venn-ABERS, followed by Platt scaling and isotonic regression.

Venn-ABERS tends to demonstrate the most significant improvement in calibration, providing notable benefits in terms of achieving well-calibrated predictions. Its utilization within the conformal prediction framework allows for reliable estimation of uncertainty and enhanced calibration performance.

To summarize, the comparison of calibration techniques revealed that Venn-ABERS tends to yield the most substantial benefits, followed by Platt scaling and isotonic regression. It is important to select the appropriate technique based on the specific requirements of the problem at hand, considering factors such as the complexity of the calibration curve and the desired level of calibration improvement.

The study's findings emphasized that miscalibrated models often exhibit overconfidence, and even logistic regression, although more calibrated than other models, can display systematic optimism. This underlines the necessity of assessing and enhancing the calibration of models to avoid unwarranted confidence and make informed decisions based on accurate probabilistic predictions. The research further indicated that uncalibrated models frequently exhibit overconfidence. This includes logistic regression, which tends to be systematically optimistic, although to a lesser extent.

Open source tools for conformal prediction in classification problems

While deep-diving into the intricacies of conformal prediction for classification, it has become evident that the right tools can significantly enhance our implementation efficiency. Recognizing this, the open source community has made remarkable contributions by providing various tools tailored for this purpose. In this section, we will explore some of the prominent open source tools for conformal prediction that can seamlessly integrate into your projects and elevate your predictive capabilities.

Nonconformist

nonconformist (https://github.com/donlnz/nonconformist) is a classical conformal prediction package that can be used for conformal prediction in classification problems.

Let's illustrate how to create an ICP using nonconformist. You can find the Jupyter notebook containing the relevant code at https://github.com/PacktPublishing/Practical-Guide-to-Applied-Conformal-Prediction/blob/main/Chapter_06.ipynb:

1. You can find the nonconformist documentation here: http://donlnz.github.io/nonconformist/index.html. First, we will install nonconformist using the standard functionality – that is, pip install:

```
!pip install nonconformist
```

2. We can import the relevant modules as follows:

```
from nonconformist.base import ClassifierAdapter
from nonconformist.cp import IcpClassifier
from nonconformist.nc import NcFactory
from nonconformist.nc import ClassifierNc,
InverseProbabilityErrFunc, andMarginErrFunc specify the
nonconformity measure using NcFactory.create_nc
```

In this case, we created an ICP with a margin nonconformity measure; we looked at this in previous chapters. This ICP uses logistic regression as the underlying classifier:

```
icp=IcpClassifier(ClassifierNc(ClassifierAdapter
(LogisticRegression()), MarginErrFunc()))
```

3. Then, we must fit the ICP using the proper training set and calibrate it using the calibration set:

```
icp.fit(X_train, y_train)
icp.calibrate(X_calib, y_calib)
```

4. Using the trained model, we can obtain the predicted class scores on the calibration and test sets:

```
y_pred_calib = model.predict(X_calib)
y_pred_score_calib = model.predict_proba(X_calib)
y_pred_test = model.predict(X_test)
y_pred_score_test = model.predict_proba(X_test)
```

In the notebook, we use a bank marketing dataset (https://archive.ics.uci.edu/dataset/222/bank+marketing) related to the direct marketing campaigns of a Portuguese banking institution. This dataset contains the following features:

- age (numeric)
- job: Type of job (categorical: admin, unknown, unemployed, management, housemaid, entrepreneur, student, blue-collar, self-employed, retired, technician, or services)
- marital: Marital status (categorical: married, divorced, or single; note that divorced means divorced or widowed)
- education (categorical: unknown, secondary, primary, or tertiary)
- default: Has credit in default? (binary: yes or no)
- balance: Average yearly balance, in euros (numeric)
- housing: Has a housing loan? (binary: yes or no)
- loan: Has a personal loan? (binary: yes or no)

The following features are related to the last contact of the current campaign:

- `contact`: Contact communication type (categorical: `unknown`, `telephone`, and `cellular`)
- `day`: Last contact day of the month (numeric)
- `month`: Last contact month of the year (categorical: `jan`, `feb`, `mar`, …, `nov`, `dec`)
- `duration`: Last contact duration, in seconds (numeric)

The other attributes are as follows:

- `campaign`: Number of contacts performed during this campaign and for this client (numeric; this includes the last contact)
- `pdays`: The number of days that passed by after the client was last contacted from a previous campaign (numeric; `-1` means that the client was not previously contacted)
- `previous`: The number of contacts performed before this campaign and for this client (numeric)
- `poutcome`: The outcome of the previous marketing campaign (categorical: `unknown`, `other`, `failure`, or `success`)

The output variable (the desired target) is `y` – has the client subscribed to a term deposit? (Binary: `yes`, `no`.)

The objective is to predict the `Class` target variable to indicate whether the marketing campaign was successful in terms of whether the client subscribed to a term deposit. The dataset is mildly imbalanced with ~12% of clients subscribing to a term deposit as a result of the marketing campaign.

We will use OpenML API to access and read the dataset. As discussed in previous chapters, ICP requires a separate calibration set that should not be used to train the underlying machine learning classifier. In the following code, we're creating three datasets – *the proper training dataset for the classifier, calibration datasets to calibrate the classifier using ICP,* and *the test dataset that will be used to evaluate the performance.* The dataset contains 45,211 instances; we must split it so that it has 1,000 instances for each of the training and calibration sets:

```
X_train_calib, X_test, y_train_calib, y_test = train_test_split(X, y,
test_size=1000, random_state=42, stratify=y)
X_train, X_calib, y_train, y_calib = train_test_split(X_train_calib,
y_train_calib, test_size=1000, random_state=42, stratify=y_train_
calib)
```

Now, we can build the underlying classifier using logistic regression and compute the accuracy and ROC AUC on the test set:

1. First, let's train logistic regression using standard scikit-learn functionality:

    ```
    model = LogisticRegression()
    model.fit(X_train, y_train)
    ```

2. Next, we must use the trained logistic regression classifier model to predict class labels and obtain class scores on the calibration and test sets:

```
y_pred_calib = model.predict(X_calib)
y_pred_score_calib = model.predict_proba(X_calib)
y_pred_test = model.predict(X_test)
y_pred_score_test = model.predict_proba(X_test)
```

3. Now, compute the classification accuracy and ROC AUC on the test set:

```
print('Classification accuracy on the test: {}'
.format(accuracy_score(y_test, y_pred_test)))
print('ROC AUC on the test set: {}'
.format(roc_auc_score(y_test, y_pred_score_test[:,1])))
```

So far, we have only used standard classification functionality. Now, let's build ICP using `nonconformist`:

1. First, we must create ICP classifiers by using a wrapper from `nonconformist`:

```
icp = IcpClassifier(ClassifierNc(ClassifierAdapter
(LogisticRegression()),MarginErrFunc()))
```

This code snippet constructs an ICP using logistic regression as the underlying machine learning model. Here's a breakdown of what's happening:

2. `LogisticRegression()`: This is a classifier from the scikit-learn library in Python that's used for binary classification tasks. It predicts the class score of an instance belonging to a particular class.

3. `ClassifierAdapter(LogisticRegression())`: This wraps the logistic regression model so that it's compatible with the nonconformity scorer. The adapter makes sure that the underlying classifier's methods align with the expectations of the nonconformity scorer.

4. `ClassifierNc(ClassifierAdapter(LogisticRegression()))`: Here, a nonconformity scorer is created. Nonconformity scorers, in the context of conformal prediction, are used to measure how much an instance deviates from the norm according to the training data. In this case, `ClassifierNc` is using `ClassifierAdapter` to create a scorer that measures nonconformity based on the logistic regression classifier.

5. `MarginErrFunc()`: This is the nonconformity measure we have looked at in previous chapters. In `nonconformist`, the margin error is defined as $0.5 - \hat{P}(y_i \mid x) - max_{y!=y_i} \hat{P}(y \mid x) \underline{\hspace{2cm}} 2$.

6. `IcpClassifier(ClassifierNc(ClassifierAdapter(LogisticRegression())), MarginErrFunc())`: Finally, an ICP is created. An ICP is a type of conformal predictor that separates the calibration set from the training set, thereby providing valid predictions even under distribution shift. It uses the defined nonconformity scorer (based on the logistic regression classifier) and the margin error function to make predictions.

7. Then, we must train the ICP classifier on the proper training set and calibrate it on the calibration set:

```
icp.fit(X_train, y_train)
icp.calibrate(X_calib, y_calib)
```

8. Now that we have trained the conformal predictor, we can compute predictions on the test set using the specified significance level:

```
# Produce predictions for the test set, with confidence 95%
prediction = icp.predict(X_test.values, significance=0.05)
```

Note that `nonconformist` uses classical conformal prediction and outputs prediction sets.

`nonconformist` is a Python library built on top of scikit-learn that focuses on implementing conformal prediction methods for classification tasks. It provides a comprehensive set of tools and algorithms to generate prediction intervals, estimate uncertainty, and enhance calibration in classification models. Here is an overview of the main features and capabilities of the `nonconformist` library:

- **Conformal prediction algorithms**: `nonconformist` offers various conformal prediction algorithms specifically designed for classification. These algorithms include the following:

  - **Inductive conformal classifier** (**ICC**): This constructs a nonconformity measure and a prediction region based on the training set

  - **Transductive conformal classifier** (**TCC**): This incorporates the test set into the construction of prediction regions

  - **Venn predictors**: This generates prediction intervals using nested Venn regions to control the number of false positives

  - **Random forest conformal predictor**: This utilizes a random forest model as the underlying classifier for conformal prediction

- **Model compatibility**: `nonconformist` seamlessly integrates with scikit-learn, allowing users to leverage scikit-learn's extensive collection of classifiers. It provides a wrapper class that allows scikit-learn classifiers to be used within the conformal prediction framework.

- **Calibration and uncertainty estimation**: The library includes functions for calibrating the output of conformal prediction models. These functions help refine the prediction intervals and ensure reliable estimates of the prediction uncertainty. `nonconformist` also offers tools to assess the calibration quality, such as reliability diagrams.

- **Evaluation and performance metrics**: `nonconformist` provides evaluation metrics to assess the performance of conformal prediction models. These metrics include accuracy, error rate, p-values, and efficiency measures, enabling thorough evaluation and comparison of different models.

- **Cross-validation support**: The library offers support for performing cross-validation with conformal prediction models. This enables robust evaluation and validation of the models across different folds of the dataset.

`nonconformist` is a powerful tool for applying conformal prediction techniques to classification problems. With its extensive range of algorithms, compatibility with scikit-learn, calibration and uncertainty estimation capabilities, evaluation metrics, and cross-validation support, `nonconformist` provides a comprehensive framework for implementing and evaluating conformal prediction models in classification tasks. It is a valuable resource for researchers and practitioners looking to incorporate conformal prediction into their classification projects.

Summary

In this chapter, we embarked on an enlightening exploration of conformal prediction specifically tailored to classification tasks. We began by underscoring the significance of calibration in the realm of classification, emphasizing its role in ensuring the reliability and trustworthiness of model predictions. Through our journey, we were introduced to various calibration methods, including the various approaches to conformal prediction. We observed how conformal prediction uniquely addresses the challenges of calibration, providing both a theoretical and practical edge over traditional methods.

We also delved into the nuanced realms of Venn-ABERS predictors, shedding light on their roles and implications in the calibration process.

Lastly, we underscored the invaluable contribution of the open source community in this domain. We highlighted tools such as the `nonconformist` library, which serve as essential resources for practitioners who are keen on implementing conformal prediction in their classification challenges.

As we conclude this chapter, it's evident that calibration, and more specifically conformal prediction, plays a pivotal role in enhancing the robustness and reliability of classification models. With the tools and knowledge we have at our disposal, we're well equipped to tackle classification problems with greater confidence and precision.

In the next chapter, we will cover conformal prediction for regression problems.

7
Conformal Prediction for Regression

In this chapter, we will cover conformal prediction for regression problems.

Regression is a cornerstone of machine learning, enabling us to predict continuous outcomes from given data. However, as with many predictive tasks, the predictions are never free from uncertainty. Traditional regression techniques give us a point estimate but fail to measure the uncertainty. This is where the power of conformal prediction comes into play, extending our regression models to produce well-calibrated prediction intervals.

This chapter delves deep into conformal prediction tailored specifically for regression problems. By understanding and appreciating the importance of quantifying uncertainty, we will explore how conformal prediction augments regression to provide not just a point prediction but an entire interval or even a distribution where the actual outcome will likely fall with pre-specified confidence. This is invaluable in many real-world scenarios, especially when making decisions based on predictions where stakes are high and being "approximately right" isn't good enough.

We will cover the following topics in the chapter:

- Uncertainty quantification for regression problems
- Various approaches to produce prediction intervals
- Conformal prediction for regression problems
- Building prediction intervals and predictive distributions using conformal prediction

Uncertainty quantification for regression problems

After completing this chapter, whenever you predict any continuous variable, you'll be equipped to add a layer of robustness and reliability to your predictions. Understanding and quantifying this uncertainty is crucial for several reasons:

- **Model interpretability and trust**: Uncertainty quantification helps us understand the reliability of our model predictions. By providing a range of possible outcomes, we can build trust in our model's predictions and interpret them more effectively.

- **Decision-making**: In many practical applications of regression analysis, decision-makers must rely on something other than point estimates. They often need to know the range within which the actual value will likely fall with a certain probability. This range, or prediction interval, provides crucial information about the uncertainty of the prediction and aids in risk management.

- **Model improvement**: Uncertainty can highlight the areas where the model may benefit from additional data or feature engineering. High uncertainty may indicate that the model needs help capturing the underlying relationship, suggesting the need for model revision or additional data.

- **Outlier detection**: Uncertainty quantification can also help us identify outliers or anomalies in the data. Observations associated with high predictive uncertainty may be outliers or indicate a novel situation not captured during model training.

Therefore, uncertainty quantification forms an essential part of regression problems. It provides a more holistic picture of predictive performance, allows for better risk management, and improves model trust and interpretability. Conformal prediction for regression, a framework that we will discuss in this chapter, is the approach to efficiently quantifying uncertainty in regression problems.

Understanding the types and sources of uncertainty in regression modeling

Uncertainty in regression modeling can arise from several sources and manifests in different ways. Broadly, these uncertainties can be classified into two main types – **aleatoric** and **epistemic**:

- **Aleatoric uncertainty**: This type of uncertainty is often called "inherent," "irreducible," or "random" uncertainty. It arises due to the inherent variability in the data itself, which is usually beyond our control. This uncertainty would not be eliminated if we were to collect more data or improve our measurements. Aleatoric uncertainty reflects the randomness, variability, or heterogeneity in our sample population.

- **Epistemic uncertainty**: Also known as "reducible" or "systematic" uncertainty, this type originates from the lack of knowledge about the system or process under study. It could be due to insufficient data, measurement errors, or incorrect assumptions about the underlying data distribution or model structure. Unlike aleatoric uncertainty, epistemic uncertainty can be reduced with more information or data.

The primary sources of these uncertainties in regression modeling are as follows:

- **Data uncertainty**: This includes measurement errors, missing values, and variability in the data. Data might be collected under different conditions or sources, adding more uncertainty to the dataset.

- **Model uncertainty**: This comes from the model's inability to precisely capture the proper relationship between predictors and the outcome. Every model makes certain assumptions (for example, linearity, independence, normality, and so on), and any violation of these assumptions introduces uncertainty.

- **Parameter uncertainty**: Every regression model involves estimating parameters (for example, coefficients). There's always some uncertainty about these estimates, which can contribute to the overall uncertainty in predictions.

- **Structural uncertainty**: This refers to uncertainty due to the choice of a specific model form. Different model structures or types (for example, linear regression, polynomial regression, and so on) might lead to different interpretations and predictions.

- **Residual uncertainty**: This comes from the residuals or errors of the model. It represents the difference between the observed outcomes and the outcomes predicted by the model.

In the context of regression modeling, recognizing these types and sources of uncertainty can help interpret the model's predictions more accurately. It can guide the process of model refinement and validation.

The concept of prediction intervals

A prediction interval is an interval estimate associated with a regression prediction, indicating a range within which the actual outcome will likely fall with a certain probability. While a point prediction gives us a singular value as the most likely outcome, a prediction interval offers a range, providing a clearer picture of the uncertainty associated with that prediction.

Why do we need prediction intervals?

Let's take a look:

- **Quantification of uncertainty**: The primary reason for employing prediction intervals is to quantify our predictions' uncertainty. No matter how sophisticated the model is, every prediction comes with inherent variability. By using prediction intervals, we can communicate this variability effectively.

- **Risk management**: In various industries, particularly finance, healthcare, and engineering, understanding the range of potential outcomes is crucial for risk assessment and mitigation. A prediction interval helps decision-makers weigh their actions' potential risks and benefits.

- **Model transparency**: Providing an interval instead of just a point estimate can enhance the transparency of the model. Stakeholders can gauge not only what the model predicts but also the confidence the model has in that prediction.

- **Guided decision-making**: Decision-makers can act more decisively when they understand the worst-case and best-case scenarios. For example, knowing the lower and upper bounds of predicted sales in sales forecasting can help allocate resources.

Understanding the necessity for prediction intervals in various contexts paves the way for a deeper discussion of their nature, particularly in comparision with confidence intervals.

How is it different from a confidence interval?

This distinction is crucial. A confidence interval pertains to the uncertainty regarding a population parameter based on a sample statistic. For instance, we might use a confidence interval to estimate the mean value of a population based on a sample mean. On the other hand, a prediction interval is about predicting a single future observation and quantifying the uncertainty around that individual prediction.

The components of a prediction interval

A prediction interval typically has the following components:

- **Lower bound**: The minimum value within the predicted range.

- **Upper bound**: The maximum value within the expected range.

- **Coverage probability**: The probability of the actual outcome falling within the prediction interval. Commonly used probabilities are 90%, 95%, and 99%.

There are various approaches to producing prediction intervals. Quantifying uncertainty in regression models is crucial for understanding the reliability of predictions. Here are some of the most used techniques:

- **Confidence and prediction intervals**: These are fundamental techniques for quantifying uncertainty. Confidence intervals provide a range of values where we expect the true regression parameters to fall, given a certain confidence level. Prediction intervals, on the other hand, give a range for predicting a new observation, incorporating both the uncertainty in the estimate of the mean function and the randomness of the new observation.

- **Resampling methods**: Techniques such as bootstrapping and cross-validation can provide empirical estimates of model uncertainty. Bootstrapping, for example, involves repeatedly sampling from the dataset with replacement and recalculating the regression estimates to get an empirical distribution of the estimates.

- **Bayesian methods**: Bayesian regression analysis provides a probabilistic framework for quantifying uncertainty. Instead of single-point estimates, Bayesian regression gives a posterior distribution for the model parameters, which can be used to construct prediction intervals.

- **Conformal prediction**: Conformal prediction is a more recent approach that measures the certainty of predictions made by machine learning algorithms. It builds prediction regions that attain valid coverage in finite samples without making assumptions about data distribution.

- **Quantile regression**: Unlike standard regression techniques, which model the conditional mean of the response variable given specific values of predictor variables, quantile regression models the conditional median or other quantiles. It can provide a more comprehensive view of the possible outcomes and their associated probabilities.

- **Monte Carlo methods**: Monte Carlo methods are a class of computational algorithms that use random sampling to obtain numerical results. In the context of uncertainty quantification, Monte Carlo methods can be used to propagate the uncertainties from the input variables to the response variable.

- **Sensitivity analysis**: Sensitivity analysis is a technique that's used to determine how different values of an independent variable will impact a particular dependent variable under a given set of assumptions. This technique is used within specific boundaries that depend on one or more input variables.

Understanding the inherent value and significance of prediction intervals makes it vital to discern the methodologies and tools to help us generate them. While traditional statistical methods have their merits, the dynamic landscape of data-driven industries necessitates more adaptive and reliable techniques. Conformal prediction, with its roots grounded in algorithmic randomness and validity, offers an enticing approach. As we transition into the next section, we will explore how conformal prediction tailors itself to regression problems, ensuring that our prediction intervals are accurate and theoretically sound. Let's dive in and unveil the intricacies of conformal prediction in the context of regression.

Conformal prediction for regression problems

In the preceding chapters, we investigated the numerous advantages that conformal prediction provides. These include the following:

- **Validity and calibration**: Conformal prediction maintains its validity and calibration, irrespective of the dataset's size. This makes it a robust method for prediction across different dataset sizes.

- **Distribution-free nature**: One of the significant benefits of conformal prediction is its distribution-free nature. It makes no specific assumptions about the underlying data distribution, making it a flexible and versatile tool for many prediction problems.

- **Compatibility with various predictors**: Conformal prediction can seamlessly integrate with any point predictor, irrespective of its nature. This property enhances its adaptability and widens its scope of application in diverse domains.

- **Non-intrusiveness**: The conformal prediction framework is non-intrusive, implying that it does not interfere with or alter the original prediction model. Instead, it is an additional layer that quantifies uncertainty, providing a holistic perspective of the model's predictions.

In regression analysis, one paramount concern is the validity and calibration of various uncertainty quantification methods. These qualities are particularly essential in generating prediction intervals, where it's expected that these intervals provide coverage that matches the specified confidence level, which can sometimes be challenging.

In the comprehensive study titled *Valid prediction intervals for regression problems*, by Nicolas Dewolf, Bernard De Baets, and Willem Waegeman (https://arxiv.org/abs/2107.00363), the authors explored four broad categories of methods designed for estimating prediction intervals in regression scenarios – Bayesian methods, ensemble methods, direct estimation methods, and conformal prediction methods:

- **Bayesian methods**: This category encompasses techniques that use Bayes' theorem to predict the posterior probability of the intervals. These methods can give robust prediction intervals by modeling the entire output distribution.

- **Ensemble methods**: Ensemble methods such as random forest or bagging can generate prediction intervals by leveraging the variability among individual models in the ensemble.

- **Direct estimation methods**: These techniques involve the direct calculation of prediction intervals. They often necessitate specific assumptions about the underlying data or the error distribution.

- **Conformal prediction methods**: Conformal prediction stands out for its distribution-free nature and ability to provide valid prediction intervals across diverse scenarios.

The authors of this study underscore that the adoption of artificial intelligence systems by humans is significantly tied to the reliability these systems can offer. The reliability here refers not only to producing accurate point predictions but also to the system's ability to gauge and communicate its uncertainty level accurately. Thus, these systems should be adept at highlighting their areas of knowledge and limitations, particularly the aspects they need clarification on.

Converting uncertainties or "what they do not know" becomes even more crucial in real-world applications, where predictions often drive significant decisions. Therefore, uncertainty quantification aids in making informed and risk-aware decisions, contributing to the broader acceptance and trust in artificial intelligence systems.

Calibration refers to the degree to which the actual coverage of a prediction interval matches its nominal coverage level. In other words, a prediction interval is well-calibrated if it contains the true value of the response variable with the expected frequency. Calibration is essential because it ensures that the prediction intervals are not too narrow or too broad and that they provide accurate information about the uncertainty associated with the predictions. In this study, the authors use conformal prediction as a general calibration procedure to ensure the prediction intervals are well-calibrated.

Methods other than conformal prediction can suffer from calibration issues because they may make assumptions about the distribution of the errors or the model parameters that do not hold in practice. For example, Bayesian methods might assume that the errors are normally distributed with a fixed variance, which may not be the case. Ensemble methods, on the other hand, may not consider the correlation between the predictions of the individual models in the ensemble. Outliers or other noise sources in the data may also affect direct estimation methods. In addition, some interval estimators can, at best, be asymptotically valid; since this is only guaranteed for infinitely large datasets, there is no guarantee that it will hold for real datasets of final size, especially for medium-sized and smaller datasets.

These issues can lead to prediction intervals that are poorly calibrated, meaning that they do not provide accurate information about the uncertainty associated with the predictions.

Conversely, conformal prediction is a non-parametric method that does not make any assumptions about the distribution of the errors or the model parameters. Instead, it uses the data to construct prediction intervals that are guaranteed to be well-calibrated, regardless of the underlying distribution of the errors.

The following table provides a summary of the characteristics of the four classes of methods for constructing prediction intervals in a regression setting:

Method	Marginal Validity	Scalability	Domain Knowledge	Validation Set
Bayesian methods	No	Only scalable with approximate inference	Yes	No
Ensemble methods	No	Yes (when scalable models are used)	No	No
Direct interval estimation	No	Yes	No	Yes
Conformal prediction	Yes	Yes (for ICP)	No	Yes

Table 7.1 – Summary of the uncertainty quantification methods

For each class of methods, this table indicates whether the method has marginal validity (meaning that it does not require any assumptions about the distribution of the errors or the model parameters), whether it is scalable (meaning that it can be applied to large datasets), whether it requires domain knowledge (meaning that it requires knowledge of the specific problem domain), and whether it requires a validation set (meaning that it requires a separate dataset to evaluate the performance of the method).

This table shows the following:

- Bayesian methods do not have marginal validity, are only scalable with approximate inference, require domain knowledge, and do not require a validation set

- Ensemble methods do not have marginal validity, are scalable, do not require domain knowledge, and do not require a validation set

- Direct interval estimation methods do not have marginal validity, are scalable, do not require domain knowledge, and require a validation set

- Conformal prediction has marginal validity, is scalable, does not require domain knowledge, and requires a validation set (in the **inductive conformal prediction (ICP)** version)

Several types of prediction interval estimators for regression problems were reviewed and compared, as follows:

- **Bayesian methods**: Gaussian process and approximate GP

- **Ensemble methods**: Dropout ensemble, deep ensembles, and mean-variance estimator

- **Direct interval estimation methods**: Neural network quantile regression

- **Conformal prediction**: Neural networks and random forest

This comparison was based on two main properties – the *coverage degree* and the *average width of the prediction intervals*.

Before calibration, the performance of prediction interval estimators varied significantly across different benchmark datasets, with large fluctuations in performance from one dataset to another. This is due to the violation of certain assumptions that are inherent to some classes of methods. For example, some methods may perform well on datasets with normally distributed errors, but poorly on datasets with strongly skewed errors. Similarly, some methods may perform well on datasets with low levels of noise, but poorly on datasets with high levels of noise.

The paper also found that the performance of the different methods for constructing prediction intervals depends on several factors, such as the size of the dataset, the complexity of the model, and the degree of skewness in the data.

For example, the paper found that Bayesian methods tend to perform better on small datasets, while ensemble methods tend to perform well on large datasets. The paper also found that the mean-variance estimator, which is a type of ensemble method, can be sensitive to the normality assumption and may perform poorly on strongly skewed datasets.

Finally, the paper found that direct interval estimation methods, such as neural network quantile regression, can be computationally expensive and may require many training samples to achieve good performance.

The conformal prediction framework was used for post-hoc calibration, and it was found that all methods attained the desired coverage after calibration, and in certain cases, the calibrated models even produced intervals with a smaller average width.

The authors illustrated how conformal prediction can be used as a general calibration procedure for methods that deliver poor results without a calibration step and show that it can improve the performance of these methods on a wide range of datasets, making conformal prediction a promising framework for constructing prediction intervals in a regression setting. In particular, the paper showed that conformal prediction can be used as a general calibration procedure for methods that deliver poor results without a calibration step. The paper also found that conformal prediction has several significant advantages over other methods for constructing prediction intervals. For example, conformal prediction has marginal validity, meaning that it does not require any assumptions about the distribution of the errors or the model parameters. Additionally, conformal prediction is scalable, meaning that it can be applied to large datasets. Finally, conformal prediction does not require domain knowledge, meaning that it does not require knowledge of the specific problem domain.

Conformal prediction is a powerful framework for quantifying uncertainty in machine learning predictions. Its principles can be applied to various types of problems involving uncertainty quantification, including regression, which involves predicting a continuous output variable.

When applied to regression problems, conformal prediction provides prediction intervals rather than point predictions. These prediction intervals offer a range of plausible values for the target variable, with a specified confidence level. For example, a 95% prediction interval indicates that we can be 95% confident that the true value of the target variable falls within this range.

The most compelling feature of conformal prediction in regression scenarios is its validity, which refers to the fact that if we claim a 95% confidence level, the true value will indeed be in the prediction interval 95% of the time. Importantly, this validity is assured for unseen test data, any sample size (not just for large or infinite samples), and any underlying point regressor model.

Moreover, conformal prediction for regression is non-parametric, which means it does not make any specific assumptions about the underlying data distribution. This makes it broadly applicable across different types of regression problems and datasets.

To implement conformal prediction for regression, we need a nonconformity measure, which quantifies how much each observation deviates from the norm. Common choices for regression include the absolute residuals. Once we have the nonconformity scores, we can generate prediction intervals based on the ordered nonconformity scores of the calibration set.

Different variants of conformal prediction can be used in regression, such as transductive (full) conformal prediction, ICP, which is more computationally efficient than classical TCP, jackknife+, and cross-conformal methods, which offer improved precision and robustness.

Overall, conformal prediction provides a flexible, robust, and reliable approach to uncertainty quantification in regression problems, offering valid and well-calibrated prediction intervals.

Building prediction intervals and predictive distributions using conformal prediction

ICP is a computationally efficient variant of the original transductive conformal prediction framework. Like all other models from the conformal prediction family, ICP is model-agnostic in terms of the underlying point prediction model and data distribution and comes with automatic validity guarantees for final samples of any size.

The key advantage of ICP compared to the original variant of conformal prediction (transductive conformal prediction) is that ICP requires training the underlying regression model only once, leading to efficient computations during the calibration and prediction phases. ICP is highly computationally efficient as the conformal layer requires very little additional computation overhead compared to training the underlying model.

The ICP process involves splitting the dataset into a proper training set and a calibration set. The training set is used to create the initial point prediction model, while the calibration set is utilized to calculate conformity scores and produce the prediction intervals of the unseen points.

ICP automatically guarantees validity and ensures that prediction intervals include actual test points with a specified degree of selected coverage.

ICP's efficiency and flexibility have made it a popular choice for uncertainty estimation in various applications.

We will use the notebook at `https://github.com/PacktPublishing/Practical-Guide-to-Applied-Conformal-Prediction/blob/main/Chapter_07.ipynb` to illustrate how to use ICP:

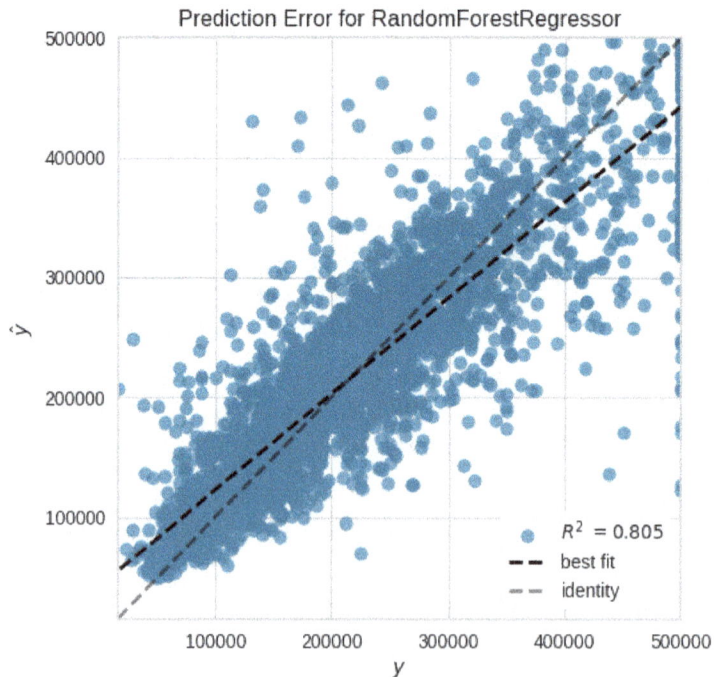

Figure 7.1 – Predictions by RandomForestRegressor on the test set

Let's generate prediction intervals by utilizing `RandomForestRegressor` as the core predictive model, and employ ICP to transform the point forecasts made by the base machine learning model into well-calibrated prediction intervals:

1. **Model training**: Initiate the process by training the chosen model using the proper training dataset:

    ```
    model = RandomForestRegressor(n_jobs=-1)
    model.fit(X_proper_train, y_proper_train)
    y_pred_calib = model.predict(X_calib)
    ```

2. **Making predictions**: Utilize the trained model to generate predictions on the calibration and test datasets:

    ```
    y_pred_calib = model.predict(X_calib)
    y_pred_test = model.predict(X_test)
    ```

3. **Nonconformity metric calculation**: For every observation within the calibration set, compute the nonconformity metric:

    ```
    y_calib_error = np.abs(y_calib - y_pred_calib)
    ```

4. **Quantile calculation**: Determine the quantile of the nonconformity metrics using the formula designed for final sample correction. The formula includes the final sample correction factor:

    ```
    q_yhat_calib = np.quantile(y_calib_error,np.ceil((n_calib+1)*(1-
    alpha))/n_calib)
    ```

5. **Form prediction intervals**: Using the quantile you calculated in the previous step, establish prediction intervals for the test set. These intervals should be based on the point predictions made by the underlying model:

    ```
    y_hat_test_lower = y_pred_test - q_yhat_calib
    y_hat_test_upper = y_pred_test + q_yhat_calib
    ```

We can illustrate the prediction outcomes on a plot that encompasses point estimates, actual values, and prediction intervals yielded by ICP:

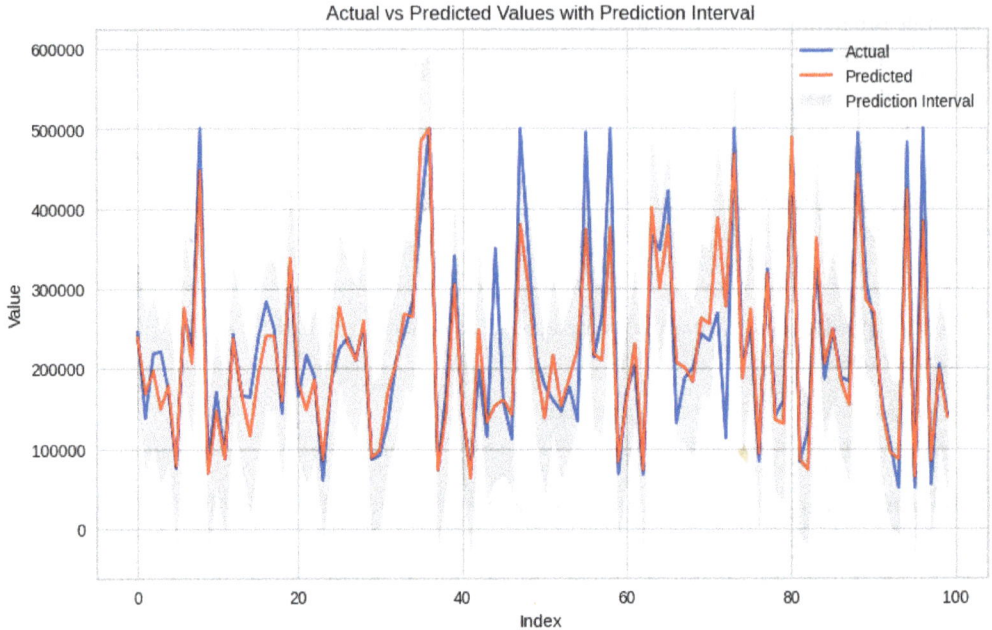

Figure 7.2 – Actual versus predicted values with a prediction interval

We can confirm that in ICP, the width of the prediction interval remains constant (as designed).

ICP for regression offers several advantages and a few drawbacks. Let's consider these.

We'll cover the advantages first:

- **Model-agnostic**: ICP can be applied to any existing regression model. This means it's a flexible method that can be used to enhance a wide variety of regression models.

- **Efficiency**: ICP only requires the underlying regression model to be trained once, leading to efficient computations during the calibration and prediction phases. This makes it computationally more efficient compared to the original transductive conformal prediction framework, which requires the model to be retrained for every new prediction.

- **Validity**: ICP comes with automatic validity guarantees. If the data distribution is exchangeable (that is, the order of the data points does not matter), then the prediction intervals produced by ICP will have the desired coverage probability.

- **Distribution-free**: ICP does not make any assumptions about the distribution of the data. This means it can be applied even when the data does not follow any known statistical distribution.

Now, let's cover the disadvantages:

- **Performance dependency**: The effectiveness of ICP relies heavily on the performance of the underlying regression model. If the underlying model does not fit the data well, the prediction intervals produced by ICP may be too wide.

- **Assumption of exchangeability**: ICP's validity guarantees depend on the assumption of exchangeability, which may not hold in many real-world scenarios (for example, when dealing with time series data).

- **Lack of adaptivity**: Unlike other conformal prediction methods, such as **conformalized quantile regression** (**CQR**) and jackknife+ methods, which we'll cover later in this chapter, ICP isn't inherently adaptive. It does not dynamically adjust to the data's complexity or structure. For example, it won't naturally produce narrower intervals in regions of the data where the model is more confident and wider intervals where the model is less confident.

In the ensuing section, we'll deep dive into one of the most popular conformal prediction models – CQR. This is a sophisticated technique that amalgamates conformal prediction's robustness with the precision of quantile regression. This fusion facilitates the generation of reliable prediction intervals and ensures that these intervals are optimally tuned to encapsulate the true values with a high degree of certainty. By harnessing the strengths of both conformal prediction and quantile regression, CQR emerges as a formidable tool for constructing well-calibrated prediction intervals, thereby augmenting the interpretability and trustworthiness of predictive models. As we traverse this section, we will unravel the mechanics of CQR, elucidate its advantages, and illustrate its application in real-world predictive scenarios.

Mechanics of CQR

In the previous section, we observed that ICP generates prediction intervals of uniform width. Consequently, it doesn't adjust adaptively to heteroscedastic data, where the variability of the response variable isn't constant across different regions of the data.

In many cases, not only it is crucial to ensure valid coverage in final samples but it is also beneficial to generate the most concise prediction intervals for each point within the input space. This helps maintain the informativeness of these intervals. When dealing with heteroscedastic data, the model should be capable of adjusting the length of prediction intervals to match the local variability associated with each point in the feature space.

CQR (developed by Yaniv Romano, Evan Patterson, and Emmanuel Candes and published in the paper *Conformalized Quantile Regression* (https://arxiv.org/abs/1905.03222)) is one of the most popular and widely adopted conformal prediction models. It was specifically designed to address the need for adaptivity by employing quantile regression as the underlying regression model.

Roger Koenker developed quantile regression – a statistical method that estimates the conditional quantiles of a response variable given a set of features. Unlike classical regression, which focuses on estimating the conditional mean, quantile regression provides a more complete picture of the relationship between the predictors and the response by estimating specified quantiles. Quantile regression provides adaptivity by allowing quantiles to adjust to the local variability of the data. This is particularly important when the data is heteroscedastic, meaning that the variance of the response variable changes across the range of the predictors. However, unlike models in the conformal prediction framework, quantile regression does not have automatic validity guarantees.

By combining the concept of quantile regression with the conformal prediction framework, CQR inherits the distribution-free validity guarantees for finite samples from conformal prediction, as well as the statistical efficiency and adaptivity of quantile regression.

Combining conformal prediction with quantile regression to create CQR provides several advantages over existing methods:

- Conformal prediction is a technique for constructing prediction intervals that attain valid coverage in finite samples, without making distributional assumptions. This means that the prediction intervals produced by CQR are guaranteed to contain the true response value with a certain probability, regardless of the underlying distribution of the data.

- Quantile regression provides a flexible and efficient way to estimate the conditional quantiles of the response variable, which allows CQR to adjust the length of the prediction intervals to the local variability of the data. This adaptivity is particularly important when the data is heteroscedastic, meaning that the variance of the response variable changes across the range of the predictors. By estimating the conditional quantiles at each query point in predictor space, CQR can construct prediction intervals that are shorter and more informative than those obtained from classical regression methods.

- According to the results published in the paper *Conformalized Quantile Regression*, CQR tends to produce shorter intervals than ICP and is adaptive. This is because CQR can adjust the length of the prediction intervals to the local variability of the data using quantile regression.

We will now describe the mechanics of CRQ, starting with classical quantile regression.

Quantile regression

Quantile regression is a statistical method that estimates the conditional quantiles of a response variable given a set of predictor variables (features). Unlike classical regression, which focuses on estimating the conditional mean, quantile regression provides a more complete picture of the relationship between the predictors and the response by estimating the entire conditional distribution.

At a high level, quantile regression minimizes a loss function that measures the difference between the observed response values and the predicted quantiles. The loss function that's used in quantile regression is typically pinball loss, a piecewise linear function that places more weight on the residuals at the specified quantile level:

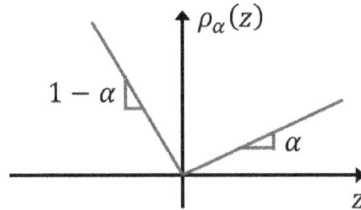

Figure 7.3 – Visualization of the pinball loss function, where y=y_hat

Under the hood, quantile regression can be implemented using a variety of algorithms, including linear quantile regression, neural networks, quantile random forest, and gradient boosting methods.

One common strategy for estimating uncertainty with quantile regression involves calculating the lower (q_lo) and upper (q_hi) quantiles for each X value in the test dataset, and then outputting [q_lo, q_hi] as a prediction interval. While this method can sometimes perform well and adapt to heteroscedasticity, it doesn't come with guarantees for achieving the desired coverage. Without such assurance for final samples, the outcomes of using such strategies could be catastrophic, particularly in critical applications such as healthcare, finance, and autonomous vehicles. This is confirmed by the results of the experiments in the paper, which show that prediction intervals produced by neural networks can significantly undercover actual values.

Several methods exist to provide asymptotic consistent results for quantile regression, including for related methods such as quantile random forests. However, none of these methods provide validity guarantees in final samples.

CQR

In previous chapters, we discussed how conformal prediction, in both its transductive and inductive forms, can offer validity guarantees in final samples. While both variants of conformal prediction can be applied to quantile regression, considering the prevalent use of ICP, we will focus solely on the application of ICP in the context of CQR.

The following figure from `https://arxiv.org/abs/1905.03222` presents a comparison between the prediction intervals produced by the standard ICP model (known as split-conformal), its locally adaptive variant, and CQR:

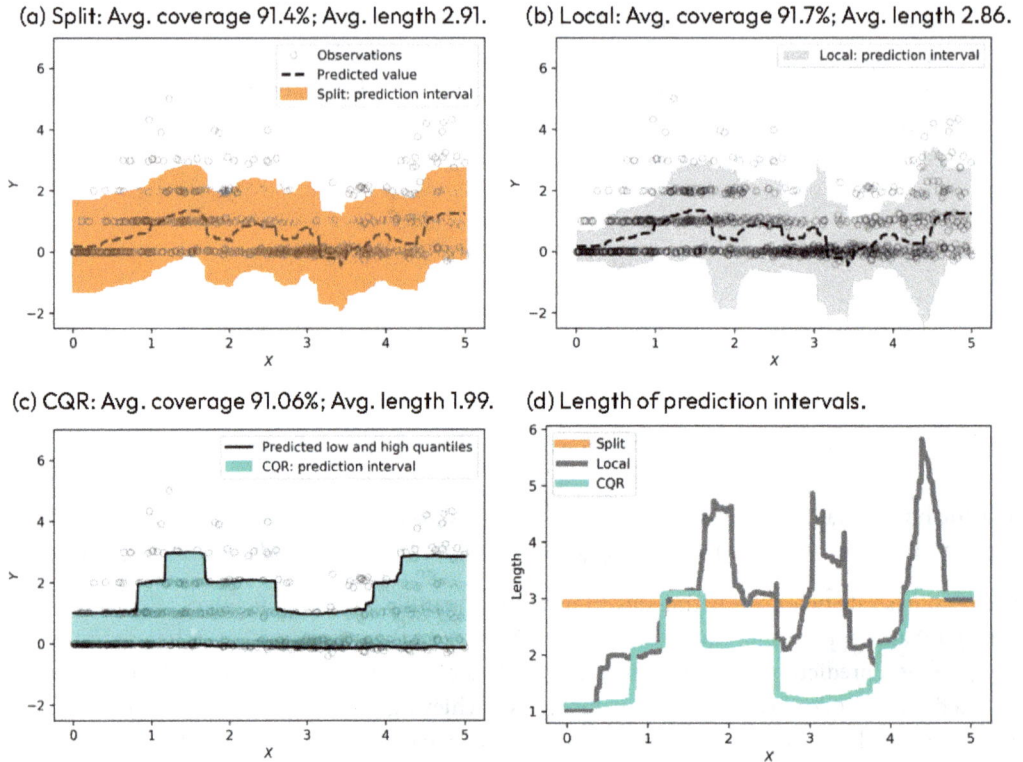

(a) Split: Avg. coverage 91.4%; Avg. length 2.91.

(b) Local: Avg. coverage 91.7%; Avg. length 2.86.

(c) CQR: Avg. coverage 91.06%; Avg. length 1.99.

(d) Length of prediction intervals.

Figure 7.4 – Prediction intervals on simulated heteroscedastic data with outliers – (a) the standard split conformal method, (b) its locally adaptive variant, and (c) CQR. The interval length as a function of X is shown in (d). The target coverage rate is 90%

The heteroscedastic dataset, which contains outliers, is modeled by three different methods, all achieving a user-specified coverage of 90%. As we discussed when we talked about ICP, ICP (split-conformal) generates prediction intervals of a constant width, indicating its non-adaptive nature. The locally weighted variant of ICP shows partial adaptivity. CQR, however, is fully adaptive and notably produces prediction intervals with the shortest average length.

Let's describe the steps involved in CQR:

1. Split the data into a proper training set and a calibration set. The proper training set is used to fit the quantile regression model, while the calibration set is used to construct the prediction intervals.

2. Fit the quantile regression model to the proper training set using any algorithm for quantile regression, such as random forest or deep neural networks. This step involves estimating the conditional quantiles of the response variable given the predictor variables.

3. Compute conformity scores that quantify the errors made by the prediction interval obtained from quantile regression [q_lo, q_hi] by computing nonconformity scores using $E_i := max$ $\left\{ \hat{q}_{\alpha_{lo}}(X_i) - Y_i, Y_i - \hat{q}_{\alpha_{hi}}(X_i) \right\}$. The nonconformity score can be interpreted as follows – if the actual label, y, falls below the lower bound of the interval, the nonconformity score corresponds to the distance from y to the lower bound. Conversely, if y exceeds the upper bound, the nonconformity score is the distance from y to the upper bound. However, if the actual label, y, is within the interval, the nonconformity score is considered negative and corresponds to the distance from y to the nearest bound.

4. Compute the prediction intervals for each test point by using empirical quantiles of E_i.

    The last step of the formula for prediction intervals is as follows:

    $$C(X_{n+1}) = \left[\hat{q}_{\alpha_{lo}}(X_{n+1}) - Q_{1-\alpha}(E, \mathcal{I}_2), \hat{q}_{\alpha_{hi}}(X_{n+1}) + Q_{1-\alpha}(E, \mathcal{I}_2) \right]$$

    Here, $Q_{1-\alpha}(E, \mathcal{I}_2) := \left(1 - \alpha \right)\left(1 + 1/|\mathcal{I}_2| \right)$ is the empirical quantile of E, $\mathcal{I}_2$ is the calibration set, and $|\mathcal{I}_2|$ is just a mathematical notation for the number of elements in the calibration set.

The conformal prediction idea here is similar to what we did in ICP – we simply use the calibration set to compute some form of nonconformity metric and then use the quantiles of the nonconformity metric to produce uncertainty intervals around the point forecast from the trained point prediction model. The only difference is that instead of adjusting the point predictions produced by the regression model, we adjust quantiles produced by the underlying statistical, machine learning, or deep learning model. Otherwise, the ideas and the mechanism of quantile calculation are the same as in ICP, given the definition of the nonconformity metric that we have described.

The key result in the paper is that if the data exchangeable, then the prediction interval constructed by the split (inductive) CQR algorithm satisfies the property of validity – that is, $\mathbb{P}\left\{ Y_{n+1} \in C(X_{n+1}) \right\} \geq 1 - \alpha$.

This is like in all other models in conformal prediction and says that given the user-specified confidence level, $1 - \alpha$, the probability of having the actual value of y within the constructed prediction interval is guaranteed to exceed $1 - \alpha$. So, if the specified user confidence is 90%, the actual points are guaranteed to lie within the prediction interval 90% of the time.

The additional bonus point is that if the nonconformity scores, E_i, are almost surely distinct, then the coverage is also bounded as $\mathbb{P}\left\{ Y_{n+1} \in C(X_{n+1}) \right\} \leq 1 - \alpha + \frac{1}{|\mathcal{I}_2| + 1}$, where $|\mathcal{I}_2|$ is the size of the calibration dataset. By way of example, for 500 points in the calibration dataset, the coverage is guaranteed to be bound between 90% and ~90.2%.

To summarize, CQR tends to produce shorter intervals compared to ICP. This is because CQR adjusts the length of the prediction intervals to the local variability of the data using quantile regression, which is particularly important when the data is heteroscedastic.

Overall, the combination of conformal prediction and quantile regression in CQR provides a powerful and flexible framework for constructing prediction intervals that are both distribution-free and adaptive to heteroscedasticity.

Jackknife+

We will now describe another widely used conformal prediction method for regression, known as jackknife+. The description of the jackknife+ technique aligns closely with the details outlined in the seminal paper *Predictive Inference with the Jackknife+*, where this method was first introduced.

We aim to fit a regression function to the training data, which consists of pairs of features (X_i, Y_i). Our goal is to predict the output, Y_{n+1}, given a new feature vector, $X_{n+1}=x$, and generate a corresponding uncertainty interval for this test point. We require that the interval contains the true Y_{n+1} given the specified target coverage – that is, $1 - \alpha : \mathbb{P}\{Y_{n+1} \in C(X_{n+1})\} \geq 1 - \alpha$.

A naive solution could be to use the residuals after fitting the underlying regressor on the training data, $|Y_i - \hat{\mu}(X_i)|$, and compute the quantile of the residuals to estimate the width of the prediction interval on the new test point, as follows:

$$\hat{\mu}(X_{n+1}) \pm \left(\text{ the } (1 - \alpha) \text{ quantile of } |Y_1 - \hat{\mu}(X_1)|, ..., |Y_n - \hat{\mu}(X_n)| \right)$$

However, in practice, such an approach will underestimate uncertainty due to overfitting as the residuals on the training set are typically smaller than residuals on unseen test data.

To circumvent overfitting, statisticians developed a robust technique known as the jackknife. This technique was originally designed to reduce bias and estimate variance. It functions by iteratively omitting one observation from the dataset and recalculating the model. This offers an empirical approach to evaluate the model's stability and resilience to individual data points:

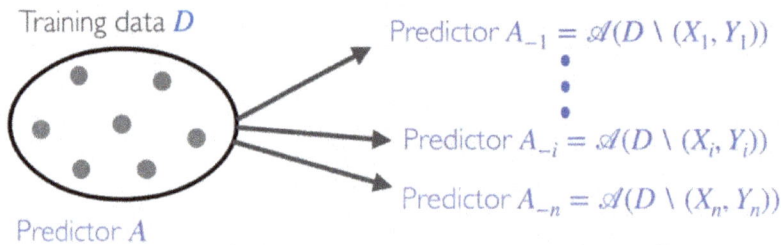

Figure 7.5 – Illustration of the jackknife prediction method

Jackknife regression

Each time jackknife regression is performed, it fits the model to all data points, excluding the one specified by (X_i, Y_i). This process allows jackknife regression to estimate the leave-one-out residual denoted by $|Y_i - \hat{\mu}_{-i}(X_i)|$. By treating these leave-one-out residuals as nonconformity scores, we can estimate the $1 - \alpha$ quantile and form prediction intervals similarly to ICP. Conceptually, this method is expected to achieve the desired coverage as it effectively tackles the issue of overfitting. This is because the residuals, $|Y_i - \hat{\mu}_{-i}(X_i)|$, are computed in an out-of-sample manner, thus providing a more realistic evaluation of model performance.

However, the jackknife procedure does not have universal theoretical guarantees and can have poor coverage properties in certain cases, particularly when the data is highly skewed or has heavy tails. Although there are statistical results under asymptotic settings or assumptions of stability regarding the jackknife regression algorithm, it is clear that in situations where the jackknife estimator is unstable, the jackknife method can result in a loss of predictive coverage.

In some cases, the jackknife method can even have zero coverage, meaning that the true value is not contained in the estimated prediction interval. Additionally, the jackknife method can be computationally intensive as it requires fitting the model multiple times on reduced datasets.

These challenges prompted the creation of an improved variant known as jackknife+. This method not only aims to bolster the original jackknife method's coverage properties and computational efficiency but also aligns with the conformal prediction family of methods. As a result, jackknife+ benefits from all the robust features of the conformal prediction framework. It guarantees validity even in final samples of any size, exhibits a distribution-free nature, and is versatile enough to be applied to any regression model.

The main difference between the jackknife and jackknife+ methods for constructing predictive intervals is that the jackknife+ method uses leave-one-out predictions at the test point to account for the variability in the fitted regression function, in addition to the quantiles of leave-one-out residuals used by the jackknife method. This modification allows the jackknife+ method to provide rigorous coverage guarantees, regardless of the distribution of the data points, for any algorithm that treats the training points symmetrically. In contrast, the original jackknife method offers no theoretical guarantees without stability assumptions and can sometimes have poor coverage properties. Leave-one-out predictions capture the uncertainty in the fitted model's predictions at the target point, allowing the intervals to adapt based on model variability. In contrast, the original jackknife method offers no theoretical guarantees without stability assumptions and can sometimes have poor coverage properties.

The jackknife+ method is related to cross-conformal prediction proposed by Vovk in that both methods aim to construct predictive intervals that provide rigorous coverage guarantees, regardless of the distribution of the data points, for any algorithm that treats the training points symmetrically.

However, the jackknife+ method differs from cross-conformal prediction in that it uses leave-one-out predictions at the test point to account for the variability in the fitted regression function and the quantiles of leave-one-out residuals used by cross-conformal prediction.

Jackknife+ regression

To recap, ICP provides validity guarantees at a user-specified $1 - \alpha$ confidence level and is highly computationally efficient as it does not necessitate retraining the base regression model. However, these advantages come at the expense of needing to divide the data into a separate calibration dataset. As this reduces the amount of data available for training the underlying regressor, it can lead to a less optimal fit for the regression model and subsequently wider prediction intervals, particularly if the original dataset is small. Conversely, if the calibration set is small, it could result in higher variability.

Jackknife+ is a modification of jackknife. Similar to jackknife, we fit the regressor n times for each of the n points in the dataset:

$$\hat{\mu}_{-i} = \mathcal{A}\left((X_1, Y_1), \ldots, (X_{i-1}, Y_{i-1}), (X_{i+1}, Y_{i+1}), \ldots, (X_n, Y_n)\right)$$

The primary distinction between the jackknife and jackknife+ methods lies in the latter's utilization of leave-one-out predictions at the test point, in addition to the quantiles of leave-one-out residuals that the jackknife method employs.

The prediction interval generated by the jackknife+ method leverages quantiles to construct prediction intervals. However, it not only examines the quantiles of leave-one-out prediction errors but it also considers the n predictions produced by the regression model for the point prediction of X_{n+1}. This approach effectively broadens the scope of the predictive model, taking into account both error estimations and individual predictions:

$$\hat{C}_{n,\alpha}^{\text{jackknife}+}(X_{n+1}) = \left[\hat{q}_{n,\alpha}^{-}\left\{\hat{\mu}_{-i}(X_{n+1}) - R_i^{\text{LOO}}\right\}, \hat{q}_{n,\alpha}^{+}\left\{\hat{\mu}_{-i}(X_{n+1}) + R_i^{\text{LOO}}\right\}\right]$$

Compare this formula with the formula for jackknife:

$$\hat{C}_{n,\alpha}^{\text{jackknife}}(X_{n+1}) = \left[\hat{q}_{n,\alpha}^{-}\left\{\hat{\mu}(X_{n+1}) - R_i^{\text{LOO}}\right\}, \hat{q}_{n,\alpha}^{+}\left\{\hat{\mu}(X_{n+1}) + R_i^{\text{LOO}}\right\}\right]$$

The significant divergence between jackknife and jackknife+ lies in the manner in which point predictions are made for the X_{n+1} test point. While the jackknife model makes the prediction only once, the jackknife+ model makes n predictions, each time fitting the model to $n-1$ data points while excluding one point at a time.

This structure effectively accommodates potential instability in the regression algorithm, an issue that previously hindered jackknife from fulfilling theoretical validity guarantees. In scenarios where the regression model is highly sensitive to the training data, the output can vary substantially. Thus, jackknife+ provides a more nuanced and flexible prediction model, catering to possible variability in the data.

The following figure presents a comparative illustration of prediction intervals produced by both the jackknife and jackknife+ models:

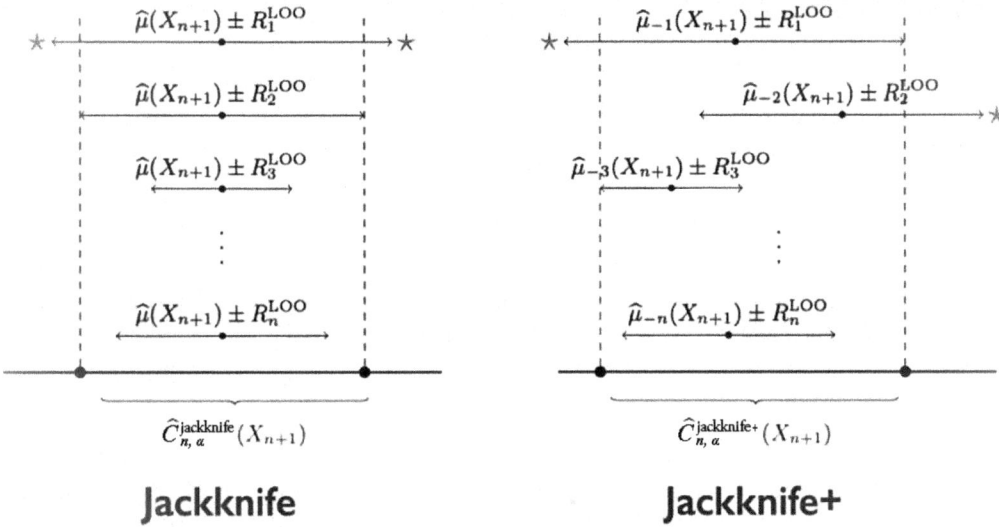

$$\widehat{\mu}(X_{n+1}) \pm R_1^{\text{LOO}}$$
$$\widehat{\mu}(X_{n+1}) \pm R_2^{\text{LOO}}$$
$$\widehat{\mu}(X_{n+1}) \pm R_3^{\text{LOO}}$$
$$\vdots$$
$$\widehat{\mu}(X_{n+1}) \pm R_n^{\text{LOO}}$$

$$\widehat{C}_{n,\,\alpha}^{\text{jackknife}}(X_{n+1})$$

Jackknife

$$\widehat{\mu}_{-1}(X_{n+1}) \pm R_1^{\text{LOO}}$$
$$\widehat{\mu}_{-2}(X_{n+1}) \pm R_2^{\text{LOO}}$$
$$\widehat{\mu}_{-3}(X_{n+1}) \pm R_3^{\text{LOO}}$$
$$\vdots$$
$$\widehat{\mu}_{-n}(X_{n+1}) \pm R_n^{\text{LOO}}$$

$$\widehat{C}_{n,\,\alpha}^{\text{jackknife+}}(X_{n+1})$$

Jackknife+

Figure 7.6 – Prediction intervals produced by jackknife and jackknife+

In scenarios where the regression algorithm is stable, both models perform quite similarly, delivering empirical coverage approximately equal to $1 - \alpha$. However, in situations where the regression model is unstable and sensitive to training data, to the extent that removing a single data point can significantly alter the predicted value at X_{n+1}, the output from both models can diverge significantly.

Contrary to the jackknife method, which lacks theoretical validity guarantees, the jackknife+ model will, even in the worst-case scenarios, assure coverage of at least 1-2\alpha. Furthermore, in most practical scenarios, barring instances involving instability, the jackknife+ model is expected to provide empirical coverage of 1-\alpha, making it a robust method for uncertainty prediction.

While the jackknife method provides a robust way of calculating nonconformity scores, it has a significant computational overhead as it requires retraining the model for each instance in the calibration set. This is where the jackknife+ method comes in.

The jackknife+ method improves upon the jackknife method by allowing for the calculation of nonconformity scores without the need to retrain the model for each instance in the calibration set. This is achieved by adjusting the nonconformity score calculation to account for the influence of each example on the model's predictions.

First, the jackknife+ method trains the model on the entire training dataset. Then, for each instance in the calibration set, the method calculates an adjusted prediction that approximates the prediction that would have been made if the instance had been left out when training the model. The nonconformity score for each instance is the absolute difference between its actual value and the adjusted prediction.

The primary benefit of the Jackknife+ method is computational efficiency as it doesn't require the model to be retrained for each instance in the calibration set.

Conformal predictive distributions

Conformal predictive distribution (CPD) is an innovative method that applies the principles of conformal prediction to generate predictive distributions. These distributions provide a comprehensive view of prediction uncertainty, offering not just interval estimates but a complete distribution over all potential outcomes.

The concept of CPD was first introduced in the paper *Nonparametric predictive distributions based on Conformal Prediction* (`https://link.springer.com/article/10.1007/s10994-018-5755-8`), by Vladimir Vovk, Jieli Shen, Valery Manokhin, and Min-ge Xie. In this paper, the authors applied conformal prediction to derive valid predictive distributions under a nonparametric assumption.

In a subsequent paper, *Conformal predictive distributions with kernels* (`https://arxiv.org/abs/1710.08894`), by Vladimir Vovk, Ilia Nouretdinov, Valery Manokhin, and Alex Gammerman, the authors reviewed the history of predictive distributions in statistics and discussed two key developments. The first was the integration of predictive distributions into machine learning, and the second was the combination of predictive distributions with kernel methods.

In the paper *Cross-conformal predictive distributions* (`http://proceedings.mlr.press/v91/vovk18a.html`), by Vladimir Vovk, Ilia Nouretdinov, Valery Manokhin, and Alexander Gammerman, the authors extended CPD to any underlying model, whether it be statistical, machine learning, or deep learning.

Inductive (split) conformal predictive systems are computationally efficient versions of conformal predictive systems that output probability distributions for labels of test observations in regression problems. These systems provide additional information that can be useful in decision-making.

Cross-conformal predictive systems are a novel application of conformal prediction that facilitates automatic decision-making. They are built on top of split conformal predictive systems, and while they can lose their validity in principle, they usually satisfy the validity requirement in practice. Therefore, cross-conformal predictive systems differ from traditional conformal predictors in that they are more computationally efficient versions that can be used for automatic decision-making.

The subsequent figure depicts the CPD prediction process for a single test instance:

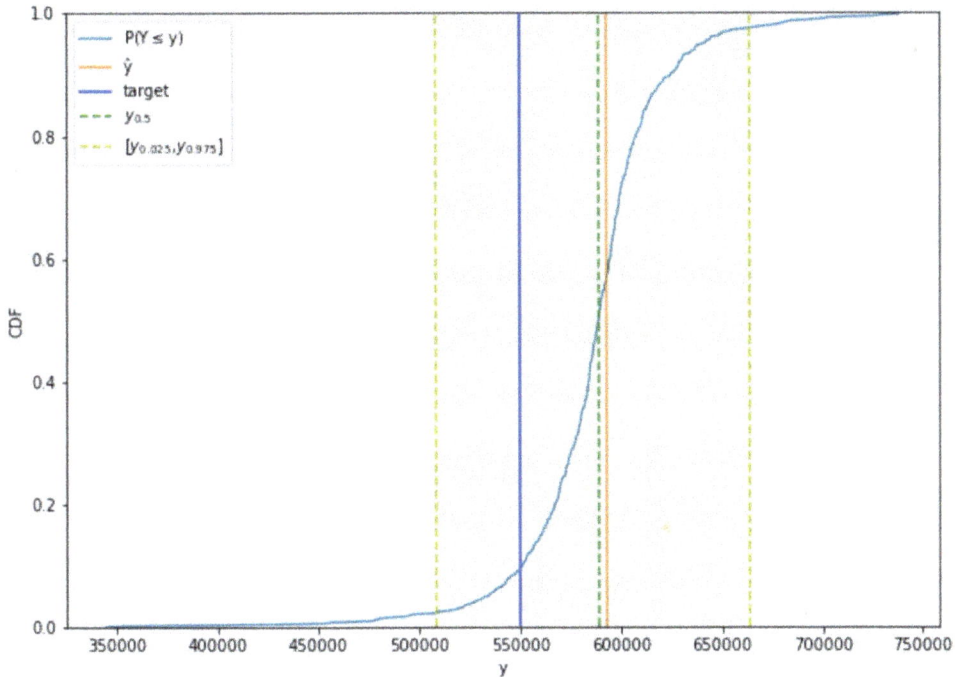

Figure 7.7 – CPD

Let's apply these concepts in practice; we will use the notebook *Conformal Prediction for Regression* (https://github.com/PacktPublishing/Practical-Guide-to-Applied-Conformal-Prediction/blob/main/Chapter_07.ipynb).

We will follow the MAPIE tutorial for CQR. The target variable of this dataset is the median house value for the California districts. This dataset comprises eight features, including variables such as the house's age, the neighborhood's median income, the average number of rooms or bedrooms, and even the location in latitude and longitude. In total, there are around 20k observations. We compute the correlation between features and also between features and the target. As is evident from the analysis, the most significant correlation of house prices is with the neighborhood's median income:

	longitude	latitude	housing_median_age	total_rooms	total_bedrooms	population	households	median_income	class
longitude	1.000000	-0.924616	-0.109357	0.045480	0.069608	0.100270	0.056513	-0.015550	-0.045398
latitude	-0.924616	1.000000	0.011899	-0.036667	-0.066983	-0.108997	-0.071774	-0.079626	-0.144638
housing_median_age	-0.109357	0.011899	1.000000	-0.360628	-0.320451	-0.295787	-0.302768	-0.118278	0.106432
total_rooms	0.045480	-0.036667	-0.360628	1.000000	0.930380	0.857281	0.918992	0.197882	0.133294
total_bedrooms	0.069608	-0.066983	-0.320451	0.930380	1.000000	0.877747	0.979728	-0.007723	0.049686
population	0.100270	-0.108997	-0.295787	0.857281	0.877747	1.000000	0.907186	0.005087	-0.025300
households	0.056513	-0.071774	-0.302768	0.918992	0.979728	0.907186	1.000000	0.013434	0.064894
median_income	-0.015550	-0.079626	-0.118278	0.197882	-0.007723	0.005087	0.013434	1.000000	0.688355
class	-0.045398	-0.144638	0.106432	0.133294	0.049686	-0.025300	0.064894	0.688355	1.000000

Figure 7.8 – California housing – correlation matrix

We can also plot the distribution of house prices:

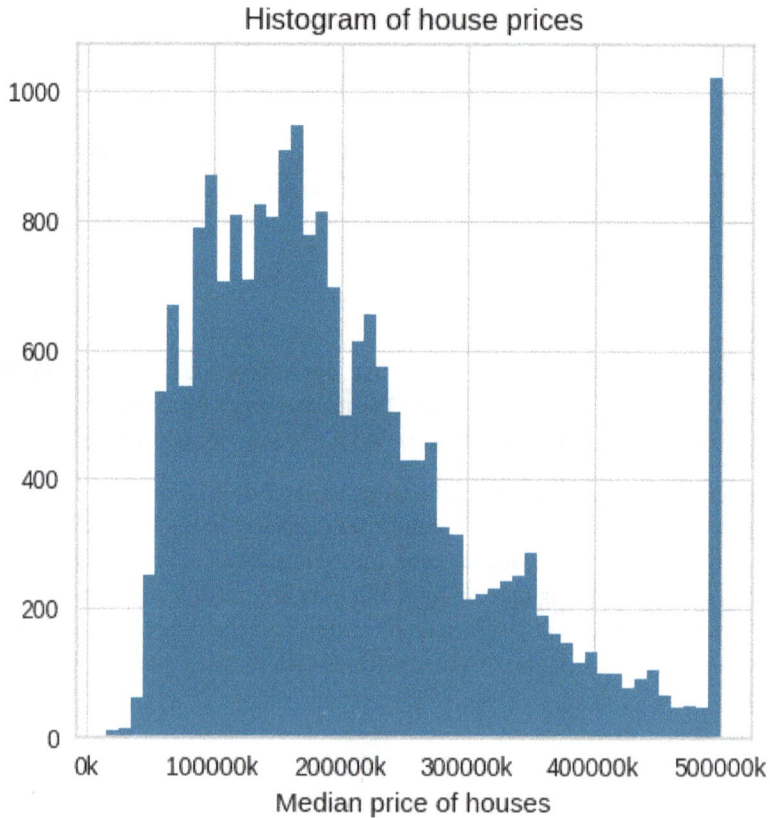

Figure 7.9 – California housing – histogram of house prices

Now, we can train and optimize the underlying model:

```
estimator = LGBMRegressor(objective='quantile', alpha=0.5, random_
state=random_state)
params_distributions = dict(
    num_leaves=randint(low=10, high=50),
    max_depth=randint(low=3, high=20),
    n_estimators=randint(low=50, high=300),
    learning_rate=uniform()
)
optim_model = RandomizedSearchCV(
    estimator,
    param_distributions=params_distributions,
    n_jobs=-1,
    n_iter=100,
    cv=KFold(n_splits=5, shuffle=True),
    verbose=0
)
optim_model.fit(X_train, y_train)
estimator = optim_model.best_estimator_
```

Several methods are used to produce probabilistic predictions, including CQR and a variant of jackknife.

We can plot the results that are produced by using different methods:

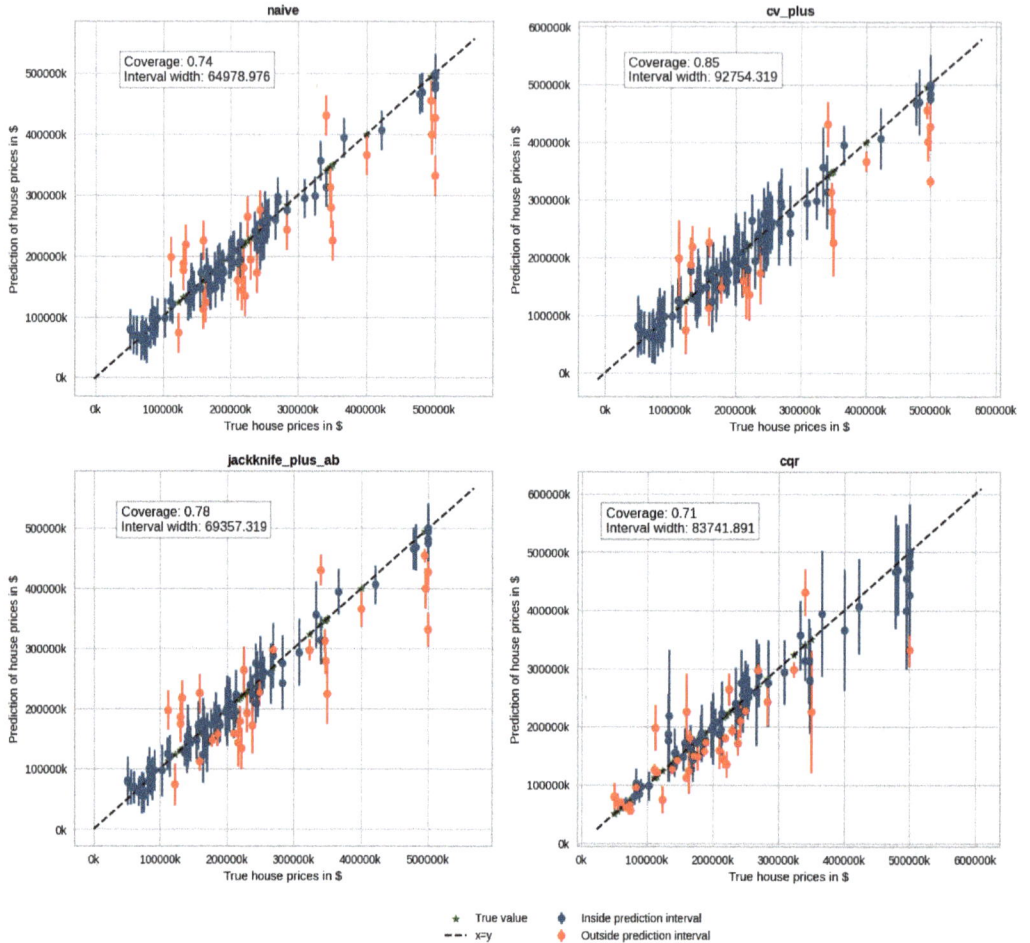

Figure 7.10 – Predicting California housing prices using various methods

We observe increased flexibility in the prediction intervals for CQR compared to other methods that maintain a fixed interval width. Specifically, as the prices rise, the prediction intervals expand correspondingly. To substantiate these observations, we will examine the conditional coverage and interval width across these intervals, segmented by quantiles:

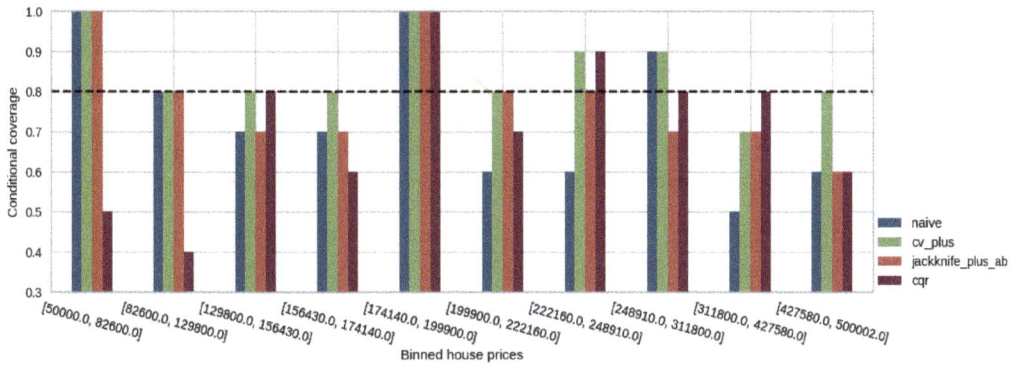

Figure 7.11 – Prediction interval coverage by binned house prices depending on price level

As we can see, CQR adjusts to larger prices more adeptly. Its conditional coverage closely aligns with the target coverage for higher prices and lower prices where other methods exceed the necessary coverage. This adaptation is likely to influence the widths of the intervals.

We can also plot interval width by bin:

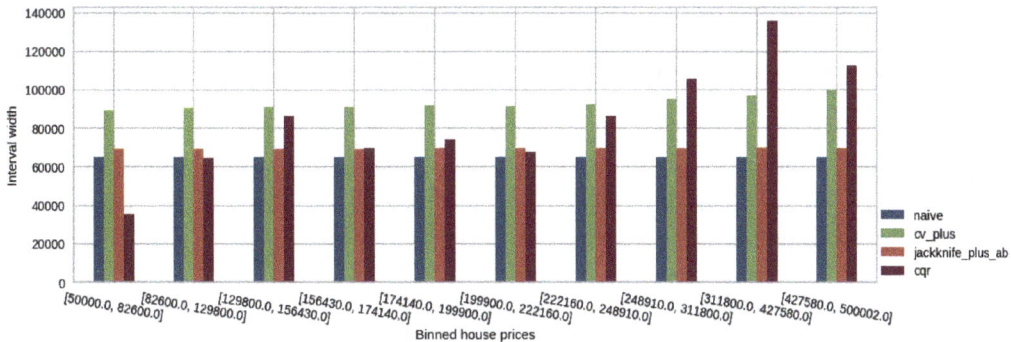

Figure 7.12 – Prediction of interval width by binned house prices depending on house price level

Now, we can look at CPD in action using the same notebook:

1. We will demonstrate the use of CPD with the Crepes package. For a quick start guide on the Crepes package, refer to https://github.com/henrikbostrom/crepes. We will wrap the standard random forest regressor using the `WrapRegressor` class from Crepes and fit it (in the usual way) to the proper training set:

   ```
   rf = WrapRegressor(RandomForestRegressor())
   rf.fit(X_prop_train, y_prop_train)
   ```

2. Then, we will calibrate the regressor using the calibration set:

   ```
   rf.calibrate(X_cal, y_cal)
   ```

3. The conformal regressor is now ready to generate prediction intervals for the test set, utilizing a 99% confidence level. The output will be a NumPy array. Each row corresponds to a test instance, with two columns indicating each prediction interval's lower and upper bounds:

   ```
   rf.predict_int(X_test, confidence=0.99)
   ```

4. We can specify that we wish to trim the intervals to omit impossible values – in this scenario, values below 0. Using the default confidence level (0.95), the resulting output intervals will be slightly more concise:

   ```
   rf.predict_int(X_test, y_min=0)
   ```

5. We will employ `DifficultyEstimator()` from `crepes` to make intervals more adaptive. Here, the difficulty is estimated by the standard deviation of the target of the default k=25 nearest neighbors in the proper training set for each object in the calibration set. First, we will obtain the difficulty estimates for the calibration set:

   ```
   de = DifficultyEstimator()
   de.fit(X_prop_train, y=y_prop_train)
   sigmas_cal = de.apply(X_cal)
   ```

6. These can now be used for the calibration, which will produce a normalized conformal regressor:

   ```
   rf.calibrate(X_cal, y_cal, sigmas=sigmas_cal)
   ```

7. We also need difficulty estimates for the test set, which we provide as input to `predict_int`:

   ```
   sigmas_test = de.apply(X_test)
   crepes_predictions = rf.predict_int(X_test, sigmas=sigmas_test,
   y_min=0)
   ```

We will get the following output:

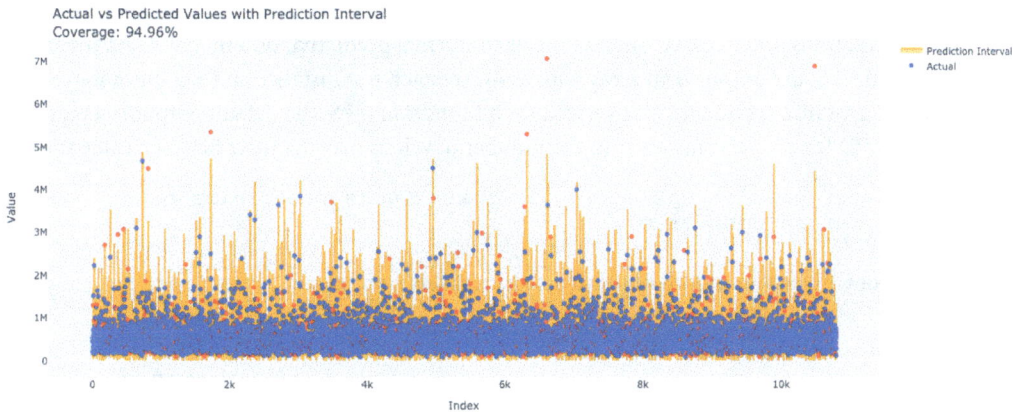

Figure 7.13 – Actual versus predicted values with prediction interval coverage

CPD (conformal predictive systems) in the Crepes package

CPD yields cumulative distribution functions (conformal predictive distributions). These not only allow us to create prediction intervals but also enable us to derive percentiles, calibrated point predictions, and p values for specific target values. Let's explore how to accomplish this.

The only modification that's required is to pass cps=True to the calibrate method:

1. For instance, we can establish normalized Mondrian conformal predictive systems by supplying both bins and sigmas to the calibrate method. In this case, we will examine Mondrian categories formed by binning the point predictions:

    ```
    bins_cal, bin_thresholds = binning(rf.predict(X_cal), bins=5)
    rf.calibrate(X_cal, y_cal, sigmas=sigmas_cal, bins=bins_cal,
    cps=True)
    ```

2. By supplying bins (and sigmas) for the test objects, we can make predictions using the conformal predictive system by utilizing the predict_cps method. This method offers flexible control over the output. In this instance, we seek to obtain prediction intervals with 95% confidence:

    ```
    bins_test = binning(rf.predict(X_test), bins=bin_thresholds)
    rf.predict_cps(X_test, sigmas=sigmas_test, bins=bins_test,
                   lower_percentiles=2.5, higher_percentiles=97.5,
    y_min=0)
    ```

3. We can instruct the `predict_cps` method to return the complete CPD for each test instance, as delineated by the threshold values, by setting `return_cpds=True`. The structure of these distributions differs based on the type of conformal predictive system. For standard and normalized CPS, the output is an array with a row for each test instance and a column for each calibration instance (residual). Conversely, for a Mondrian CPS, the default output is a vector with one CPD for each test instance, as the number of values may fluctuate between categories:

    ```
    cpds = rf.predict_cps(X_test, sigmas=sigmas_test, bins=bins_
    test, return_cpds=True)
    ```

4. We can plot CPD for a random object from the test set:

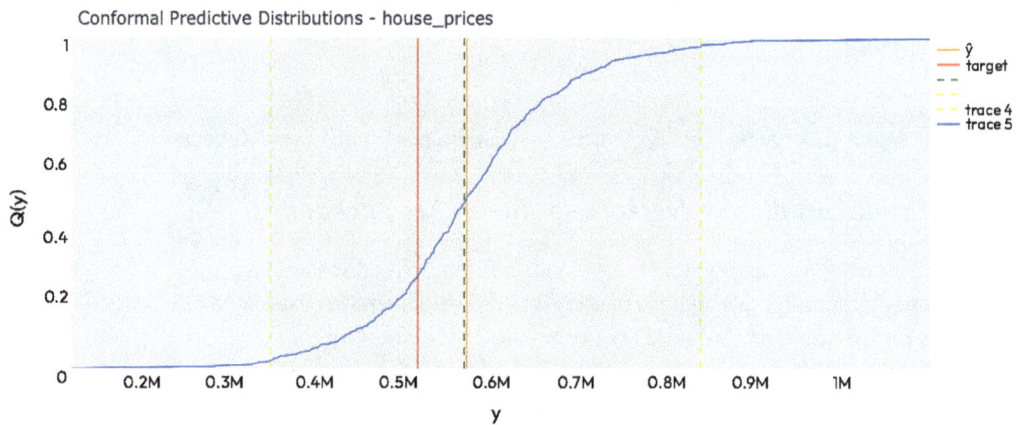

Figure 7.14 – CPD for a test object

Summary

This chapter explored uncertainty quantification for regression problems, a critical aspect of data science and machine learning. It highlighted the importance of uncertainty and the methods to handle it effectively to make more reliable predictions and decisions.

One of the significant sections of this chapter was dedicated to various approaches that can be used to produce prediction intervals. It systematically broke down and explained diverse methods, elucidating how each works and their advantages and disadvantages. This detailed analysis aids in understanding the mechanisms behind these approaches and their practical application in real-world regression problems.

Furthermore, this chapter discussed building prediction intervals and predictive distributions using conformal prediction. We provided a step-by-step guide to constructing these intervals and distributions. This chapter also offered practical insights and tips for effectively utilizing conformal prediction to achieve more reliable and trustworthy predictions in regression problems.

In addition, we delved into advanced topics such as CQR, jackknife+, and CPD. These advanced techniques were broken down and explained in detail, helping you grasp their complexity and utility in handling regression problems.

Practical illustration was crucial in this chapter, offering hands-on experience and insights. This chapter utilized housing price datasets to demonstrate the application of the discussed models and techniques. Libraries such as MAPIE and Crepes were applied to the datasets, providing you with practical knowledge and experience beyond theoretical understanding.

In conclusion, this chapter provided a comprehensive and practical guide that covered various topics related to uncertainty quantification, prediction intervals, and conformal prediction for regression problems. The realistic illustrations, using real-world datasets and libraries, further enhanced the learning experience, making this chapter a valuable resource for anyone looking to deepen their understanding and strengthen their skills in these critical areas.

8

Conformal Prediction for Time Series and Forecasting

In this chapter, we will explore the exciting field of **conformal prediction** for time series and forecasting. Conformal prediction is a powerful tool for producing **prediction intervals** (**PIs**) for point forecasting models, and we will show you how to apply this technique to your data using open source libraries. This chapter will take you on a journey from understanding the fundamentals of **uncertainty quantification** (**UQ**) in time series to the intricate mechanisms behind conformal prediction in forecasting.

With this chapter, you will have a solid understanding of the various approaches to producing PIs, and you will be able to build your PIs using conformal prediction.

In this chapter, we're going to cover the following main topics:

- UQ for time series and forecasting problems
- The concept of PIs in forecasting applications
- Various approaches to producing PIs
- Conformal prediction for time series and forecasting

By the end of this chapter, you'll be able to apply the concepts and open source tools that will be discussed to your industry applications, providing robust forecasting with well-defined uncertainty bounds. These lessons will enhance your forecasting abilities, giving your models an edge by allowing you to add confidence measures to your predictions.

UQ for time series and forecasting problems

UQ is not just a sophisticated addition to time series forecasting; it is a fundamental aspect that provides invaluable insights into the nature of the predictions. Let's look at why it's important and a brief history of its development.

The importance of UQ

UQ is a critical component of time series forecasting. While a forecast model may provide accurate predictions on average, understanding the uncertainty around those predictions is equally essential. There are several key reasons why properly quantifying uncertainty is vital for practical time series forecasting:

- **Risk assessment**: In many domains, such as finance, healthcare, and environmental science, forecasting is closely linked with decision-making. Understanding the uncertainty in predictions aids in assessing potential risks, thus enabling informed decisions.

- **Model confidence**: UQ provides an understanding of the confidence in each model's predictions. This can lead to a more refined model selection and help identify areas where the model may be underperforming.

- **Optimization of resources**: By acknowledging the uncertainty, resources can be allocated more optimally. For example, understanding the uncertainty in demand forecasts in supply chain management may lead to better inventory management.

- **Regulatory compliance**: In some industries, quantifying the uncertainty of forecasts might be mandated by regulatory bodies, emphasizing the importance of a systematic approach to UQ

Having established the critical role of UQ, we now turn to its evolution.

The history of UQ

The need to provide reliable measures of uncertainty alongside time series forecasts has long been recognized. Over the decades, advances in statistical and computational methods have enabled more sophisticated approaches to quantifying uncertainty. Some of the significant historical developments are as follows:

- **Early statistical methods**: The roots of UQ in time series can be traced back to the early statistical models. Techniques such as PIs were applied to provide bounds on predictions.

- **Bayesian approaches**: Bayesian methods brought a probabilistic perspective to UQ, allowing for more nuanced uncertainty descriptions. Bayesian forecasting models incorporate prior beliefs and likelihood functions to create posterior distributions, representing uncertainty comprehensively.

- **Bootstrapping and resampling**: Techniques such as bootstrapping enabled UQ without strong parametric assumptions, making it accessible for more complex models.

The historical developments we've explored provided a critical foundation for UQ in time series analysis. Now, let's dive deeper into some of those early statistical techniques and see how they enabled the first steps toward quantifying forecast uncertainty.

Early statistical methods – the roots of UQ in time series

UQ has always been a critical part of statistical analysis, and its role in time series forecasting is no different. The early days of statistical modeling laid the foundation for understanding uncertainty in predictions, and various techniques were developed to provide bounds on forecasts. Here, we will investigate some of these early statistical methods and see how they paved the way for the modern understanding of UQ in time series analysis.

One of the seminal contributions to UQ in time series forecasting was the concept of confidence intervals:

- **T-distribution for small samples**: When dealing with small sample sizes, the t-distribution provided more accurate intervals, accounting for the increased uncertainty due to limited data
- **Interval estimation for autoregressive models**: Specific techniques were developed for time series models such as ARIMA, where the confidence intervals could be derived for the parameters and forecasts

Along with confidence intervals, prediction bounds were developed to encapsulate the uncertainty associated with future observations. These bounds considered the uncertainty in the model parameters and the random nature of future errors:

- **Prediction error variance**: By estimating the prediction error variance, bounds could be created around the forecast values
- **Forecast error decomposition**: Techniques were developed to decompose the forecast error into various components, providing insights into the sources of uncertainty

While these early methods were highly influential, they often relied on strong assumptions about the underlying distributions and model structure. The parametric nature of these techniques made them less flexible in dealing with complex, non-linear time series data:

- **Non-parametric methods**: Recognizing these limitations led to the development of non-parametric methods that didn't rely on specific distributional assumptions
- **Robust statistical techniques**: Efforts were also made to create more robust statistical methods that could handle outliers and non-constant variance, extending the scope of early UQ methods

The early statistical methods for UQ in time series laid the groundwork for subsequent advancements in this field. The principles embedded in these techniques, such as confidence intervals and prediction bounds, continue to be central to modern UQ approaches. They represent a legacy that's been built upon and refined, leading to various current methods for understanding uncertainty in time series forecasting.

Now, let's dive deeper into some of those early statistical techniques and see how they enabled the first steps toward quantifying forecast uncertainty.

Modern machine learning approaches

The previous sections explored the early statistical foundations of UQ for time series predictions. These techniques, while pioneering, relied heavily on parametric assumptions and simple model structures. The rise of modern machine learning has enabled more flexible and robust approaches to quantifying uncertainty, overcoming some limitations of traditional methods. Let's look at the key innovations in this area:

- **Modern machine learning approaches**: With the rise of machine learning, techniques such as dropout and ensemble methods have been developed to quantify uncertainty.

- **Conformal prediction**: Recently, conformal prediction has emerged as a robust framework for UQ. It provides a non-parametric approach, guaranteeing valid PIs under mild assumptions.

UQ is integral to time series forecasting. It enriches the understanding of the predictions, facilitates better decision-making, and aligns with regulatory requirements. The evolution of UQ over time has led to diverse approaches, each adding value in different contexts.

The recent advent of conformal prediction, which will be explored later in this chapter, represents a significant advancement in this field, offering robust and universally applicable uncertainty measures.

In summary, the emergence of flexible machine learning techniques has enabled robust new approaches to UQ that overcome the limitations of early statistical methods. This evolution has provided a diverse toolkit for quantifying uncertainty in time series forecasting.

Next, we will explore the concepts behind PIs, a foundation for communicating forecast uncertainty.

The concept of PIs in forecasting applications

PIs are vital tools in forecasting, providing a range of plausible values within which a future observation is likely to occur. Unlike point forecasts, which give a single best estimate, PIs communicate the uncertainty surrounding that estimate.

This section explores the fundamental concepts behind PIs and their significance in various forecasting applications.

Definition and construction

PIs are constructed around a point forecast to represent the range within which future observations are expected to lie with a given confidence level. For example, a 95% PI implies that 95 of 100 future observations are expected to fall within the defined range.

PIs can take several forms, depending on the approach used to generate them. Two key distinguishing factors are as follows:

- **Symmetric versus asymmetric intervals**: PIs can be symmetric, where the bounds are equidistant from the point forecast, or asymmetric, reflecting differing uncertainty in different directions

- **Parametric versus non-parametric methods**: PIs can be created using parametric (for example, assuming normal distribution) or non-parametric methods, depending on the underlying data distribution assumptions

The importance of forecasting applications

PIs play an essential role in various forecasting domains, and here's why:

- **Decision making**: PIs enable decision-makers to assess risks and opportunities – for instance, they allow investors to gauge the volatility of an asset

- **Model evaluation**: Comparing actual observations with PIs can be a part of model diagnostic checks, helping evaluate a model's adequacy in capturing uncertainty

- **Optimizing operations**: In supply chain management, PIs can aid in optimizing inventory by reflecting the uncertainty in demand forecasts

- **Communicating uncertainty**: PIs effectively communicate uncertainty to non-technical stakeholders, facilitating more nuanced discussions and planning

Challenges and considerations

While highly valuable, constructing accurate and reliable PIs is not without challenges:

- **Assumption sensitivity**: PIs may be sensitive to the underlying assumptions about the data distribution, and incorrect assumptions can lead to misleading intervals.

- **Coverage and width trade-off**: Achieving the correct coverage probability (95%) often competes with the desire for narrow intervals. Wider intervals may cover the desired percentage of observations but may need to be more informative.

- **Computational complexity**: Some methods for constructing PIs can be computationally intensive, particularly with large datasets or complex models.

PIs are at the heart of UQ in forecasting applications, offering a more comprehensive view of prospects. They support strategic decision-making, enable model evaluations, and foster effective communication of uncertainty. Understanding the concept and practicalities of PIs is essential for anyone working with forecasting models, providing a means to navigate and leverage the inherent uncertainty in predicting future outcomes.

While PIs bring invaluable insights, constructing accurate and informative intervals is only sometimes straightforward, as we've seen. However, decades of research have produced diverse techniques to tackle these challenges.

Various approaches to producing PIs

PIs are an essential tool in forecasting, allowing practitioners to understand the range within which future observations are likely to fall. Various approaches have been developed to produce these intervals, each with advantages, applications, and challenges. This section will explore the most prominent techniques for creating PIs.

Parametric approaches

Parametric approaches make specific assumptions about the distribution of forecast errors to derive PIs. Some standard techniques in this category are as follows:

- **Normal distribution assumptions**: By assuming that the forecast errors follow a normal distribution, we can compute symmetric PIs based on standard errors and critical values from the normal distribution.

- **Time series models**: Models such as ARIMA and exponential smoothing can generate PIs by modeling the underlying stochastic process and using the estimated parameters to produce intervals.

- **Generalized linear models (GLMs)**: GLMs extend linear models to non-normal distributions, allowing for more flexible PI construction. GLMs broaden linear regression to response variables that follow distributions other than the normal distribution. GLMs allow us to model data with non-normal responses such as binary, count, or categorical outcomes. Like linear models, GLMs relate the mean response to explanatory variables through a link function and linear predictor. However, the response distribution can be non-normal, handled via an exponential family log-likelihood. Here are some common examples of GLMs:

  - Logistic regression for binary classification (logit link, binomial distribution)

  - Poisson regression for count data (log link, Poisson distribution)

  - Multinomial regression for categorical responses (logit link, multinomial distribution)

GLMs estimate coefficients for each feature, just like ordinary linear regression. However, by expanding the response distribution and link function, they can model non-normal processes needed for regression-style prediction with non-continuous targets.

Their flexibility makes GLMs helpful in constructing PIs for a broader range of problems compared to standard linear regression. The intervals incorporate the modeled response distribution.

Non-parametric approaches

Non-parametric methods aim to construct PIs without making strict assumptions about the distribution of forecast errors. Some fundamental techniques in this category are as follows:

- **Bootstrapping**: Bootstrapping involves resampling the observed data and estimating the distribution of forecasts, from which PIs can be derived

- **Quantile regression**: This method directly models the response variable's quantiles, enabling the construction of PIs without specific distributional assumptions

- **Empirical percentiles**: Using historical and empirical percentiles, we can construct PIs without parametric assumptions

Bayesian approaches

The Bayesian statistical framework provides a probabilistic approach to generating PIs by explicitly modeling different sources of uncertainty. Two critical techniques for Bayesian PI construction are as follows:

- **Bayesian forecasting models**: Bayesian models provide a probabilistic framework that captures uncertainty in parameters and predictions, allowing for the direct calculation of PIs from posterior distributions

- **Monte Carlo Markov Chain (MCMC) sampling**: MCMC sampling can be used to simulate the posterior distribution of a Bayesian model, enabling the construction of PIs

Machine learning approaches

The flexibility of modern machine learning models provides new opportunities for generating PIs in a data-driven manner. By leveraging techniques tailored for these highly complex and nonlinear models, valid PIs can be obtained without strict distributional assumptions. Let's look at some machine learning approaches:

- **Ensemble methods**: Techniques such as random forest and gradient boosting machines can create PIs by using the distribution of predictions from individual ensemble members

- **Neural network quantile regression**: Neural networks can be trained to predict specific quantiles, forming the basis for PIs

- **Dropout as a Bayesian approximation**: In deep learning, dropout can approximate Bayesian inference, allowing for UQ and PI construction

Conformal prediction

As a non-parametric, distribution-free framework, conformal prediction can be integrated with various modeling approaches to produce PIs. Producing PIs for time series forecasting models using conformal prediction methods is the main subject of this chapter.

Producing PIs is multifaceted, with various approaches tailored to different data types, models, and requirements. From traditional statistical methods to cutting-edge machine learning techniques and conformal prediction, the field of PI construction is rich and diverse. Understanding these approaches empowers practitioners to select the most appropriate method for their forecasting application, balancing accuracy, interpretability, computational efficiency, and other considerations. Whether operating within rigorous parametric assumptions or exploring flexible non-parametric techniques, these methods offer valuable insights into the uncertainty inherent in forecasting.

Conformal prediction, a robust framework for generating PIs for point forecasting models, has been widely utilized in time series and forecasting applications. Numerous studies have chronicled the evolution and popularity of various conformal prediction models for time series forecasting.

Conformal prediction for time series and forecasting

Creating reliable PIs for time series forecasting has been a longstanding, intricate challenge that remained unsolved for years until conformal prediction emerged.

This problem was underscored during the 2018 M4 Forecasting Competition, which necessitated participants to supply PIs and point estimates.

In the research paper titled *Combining Prediction Intervals in the M4 Competition*, (`https://www.sciencedirect.com/science/article/abs/pii/S0169207019301141`), Yael Grushka-Cockayne from the Darden School of Business and Victor Richmond R. Jose from Harvard Business School scrutinized 20 interval submissions. They assessed both the calibration and precision of the predictions and gauged their performances across different time horizons. Their analysis concluded that the submissions were ineffective in accurately estimating uncertainty.

Ensemble batch PIs (EnbPIs)

Conformal Prediction Intervals for Dynamic Time-Series (`http://proceedings.mlr.press/v139/xu21h/xu21h.pdf`), by researchers Chen Xu and Yao Xie from Georgia Tech University, was the first paper to implement conformal prediction for time series forecasting and was presented at the prestigious conference ICML in 2021.

EnbPI is currently one of the most popular implementations of conformal prediction for time series forecasting. It has been implemented in popular open source conformal prediction libraries such as MAPIE, Amazon Fortuna, and PUNCC.

The study introduces a technique for creating PIs not bound by any specific distribution for dynamic time series data. The EnbPI method encompasses a bootstrap ensemble estimator to formulate sequential PIs. Unlike classical conformal prediction methods that require data exchangeability, EnbPI does not require data exchangeability and has been custom-built for time series.

The data exchangeability assumption suggests that the sequence in which observations appear in the dataset doesn't matter. However, this assumption does not apply to time series, where the sequence of data points is crucial. EnbPI doesn't rely on data exchangeability, making it aptly suited for time series analysis.

PIs generated by EnbPI attain a finite-sample, approximately valid marginal coverage for broad regression functions and time series under the mild assumption of strongly mixing stochastic errors. Additionally, EnbPI is computationally efficient and avoids overfitting by not requiring data splitting or training multiple ensemble estimators. It is also scalable to producing arbitrarily many PIs sequentially and is well suited to a wide range of regression functions.

Time series data is dynamic and often non-stationary, meaning the statistical properties can change over time. While various regression functions exist for predicting time series, such as those using boosted trees or neural network structures, these existing methods often need help constructing accurate PIs. Typically, they can only create reliable intervals by placing restrictive assumptions on the underlying distribution of the time series, which may only sometimes be appropriate or feasible.

Here's a simplified version of the steps to build an EnbPI predictor:

1. **Select a bootstrap ensemble estimator**: Any bootstrap ensemble estimator can be used with EnbPI.

2. **Train the ensemble estimators**: The base forecasting model is trained multiple times on different bootstrap samples drawn from the original training data to generate the ensemble. Each bootstrap sample is created by sampling with replacement from the training set. This results in an ensemble of models with slightly different training data.

3. **Compute residuals**: For each point in t = 1,..., T, calculate the residuals using ensemble estimators that did not use point *t* for training. The aim is to use out-of-sample errors as a nonconformity measure to indicate the variance of predictions. All such out-of-sample errors are compiled into a single array.

4. **Generate predictions**: The ensemble estimator generates point predictions for the test data.

5. **Construct PIs**: The PIs are constructed using the predictions from the ensemble estimator and a chosen significance level. Like many other conformal prediction methods, a quantile with a specified confidence level can be applied to the distribution of out-of-sample errors created in *step 3*. This quantile value is then used to create PIs by applying the quantile value to the aggregated point prediction produced using a trained ensemble estimator.

To demonstrate EnbPI in action, we will use Amazon Fortuna (`https://aws-fortuna.readthedocs.io/en/latest/index.html`) and follow its example, *Time series regression with EnbPI, a conformal prediction method* (`https://aws-fortuna.readthedocs.io/en/latest/examples/enbpi_ts_regression.html`). You can find the Jupyter notebook, `Chapter_08_EnbPI_ipynb.ipynb`, in this book's GitHub repository: `https://github.com/PacktPublishing/Practical-Guide-to-Applied-Conformal-Prediction/blob/main/Chapter_08_EnbPI.ipynb`. Let's get started:

1. First, we will install Amazon Fortuna with `pip install`:

    ```
    !pip install aws-fortuna
    ```

2. We will use the `Bike sharing` demand dataset, available on scikit-learn:

    ```
    from sklearn.datasets import fetch_openml
    bike_sharing = fetch_openml("Bike_Sharing_Demand", version=2,
    as_frame=True, parser="pandas")
    ```

3. Let's inspect the dataset header:

season	year	month	hour	holiday	weekday	workingday	weather	temp	feel_temp	humidity	windspeed	count
spring	0	1	0	False	6	False	clear	9.84	14.395	0.81	0.0	16
spring	0	1	1	False	6	False	clear	9.02	13.635	0.80	0.0	40
spring	0	1	2	False	6	False	clear	9.02	13.635	0.80	0.0	32
spring	0	1	3	False	6	False	clear	9.84	14.395	0.75	0.0	13
spring	0	1	4	False	6	False	clear	9.84	14.395	0.75	0.0	1

Figure 8.1 – Bike sharing demand dataset

The dataset contains information about bike-sharing rentals, including additional information such as temperature, humidity, and wind speed. The problem requires forecasting bike sharing demand expressed in the count of rented bikes.

4. We can calculate the demand, grouped by weekday and hour, and illustrate the results with the following plot:

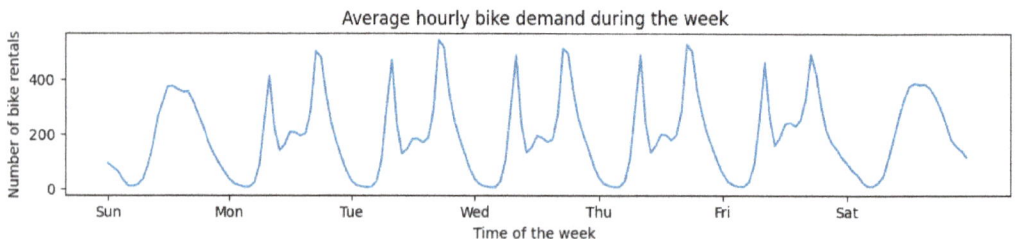

Figure 8.2 – Average hourly bike demand during the week

EnbPI requires bootstrapping the data – that is, sampling with replacement random subsets of the time series and training a model for each sample.

5. We can test the `DataFrameBootstrapper` class and look at an example of bootstrapped data samples. For example, the first bootstrapped sample looks like this:

	season	year	month	hour	holiday	weekday	workingday	weather	temp	feel_temp	humidity	windspeed
16864	winter	1	12	11	False	1	True	rain	18.86	22.725	0.94	15.0013
14012	fall	1	8	1	False	0	False	clear	25.42	29.545	0.78	8.9981
11443	summer	1	4	0	False	5	True	clear	22.14	25.760	0.68	36.9974
6325	winter	0	9	3	False	1	True	misty	25.42	27.275	0.94	0.0000
8490	spring	0	12	11	False	0	False	clear	12.30	15.150	0.70	11.0014
...	...	...	...	...	...	...	...	...	...	...	...	...
11931	summer	1	5	8	False	4	True	clear	21.32	25.000	0.55	30.0026
4606	fall	0	7	20	False	5	True	clear	28.70	32.575	0.51	15.0013
10823	summer	1	4	2	False	0	False	misty	14.76	17.425	0.76	11.0014
15921	winter	1	11	2	False	4	True	clear	13.12	16.665	0.66	8.9981
15264	winter	1	10	5	False	3	True	misty	25.42	27.275	0.94	7.0015

Figure 8.3 – Example of a bootstrapped sample

6. We can check for duplicates in this bootstrapped sample – as bootstrapping is with replacement, as expected, we can see that some objects have been duplicated during bootstrapping:

	season	year	month	hour	holiday	weekday	workingday	weather	temp	feel_temp	humidity	windspeed
8490	spring	0	12	11	False	0	False	clear	12.3	15.15	0.7	11.0014
8490	spring	0	12	11	False	0	False	clear	12.3	15.15	0.7	11.0014

Figure 8.4 – Duplicated objects in the bootstrapped sample

7. We can now train the model for each bootstrapped sample. To evaluate conformal PIs, we can calculate the coverage probability, which measures the percentage of test observations that fall within the generated intervals, and check what proportion of intervals contain the point predictions.

8. Ultimately, we evaluate the dimension of the conformal intervals, which, in this scenario where no online feedback is given, are assumed by EnbPI to be uniform for all intervals.

    The percentage of intervals containing actual targets is 0.95, while the size of the conformal intervals is 0.4446.

Using EnbPI, we created PIs based on a user-defined coverage of 0.95. Contrary to most other UQ methods, which often fail to meet the user-specified confidence level, conformal prediction meets user requirements. It consistently generates PIs that align with the user-defined confidence level.

Let's plot the predictions:

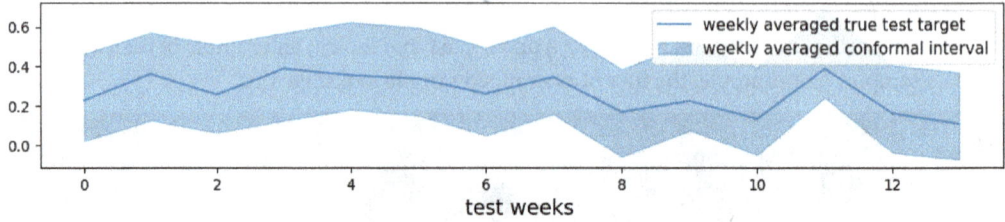

Figure 8.5 – Predictions using EnbPI

The EnbPI model is adept at avoiding overfitting, ensuring computational efficiency, and is scalable for producing numerous PIs sequentially. In practice, the EnbPI model creates PIs in line with user-defined confidence levels, ensuring reliability in its predictions. We have provided practical examples using Amazon Fortuna and the bike-sharing demand dataset from scikit-learn, demonstrating the model's capability to accurately gauge **prediction interval coverage probability** (PICP) and the size of the conformal intervals.

NeuralProphet

NeuralProphet is a forecasting framework built on PyTorch that merges the interpretability of traditional methods with the scalability of deep learning models. It's trained using standard deep learning techniques and provides accurate and interpretable results for various forecasting applications.

The framework introduces local context through auto-regression and covariate modules, which can be set up as either classical linear regression or neural networks. This allows NeuralProphet to handle short-term forecasting while capturing complex nonlinear relationships between variables. The auto-regression module models the dependency of the target variable on its past values, while the covariate module addresses its dependence on other known variables.

NeuralProphet is designed to be user-friendly, offering reliable defaults and automatic hyperparameters for beginners while allowing experienced users to input domain knowledge through optional model customizations. As a successor to Facebook Prophet, it retains the foundational components but improves precision and scalability.

The essential model components of NeuralProphet include the trend, seasonality, holidays, auto-regression, and covariate modules. These additive components can be scaled by the trend for a multiplicative effect. Each module has its inputs and modeling processes, but all modules must produce h outputs, where h is the number of steps to be forecasted into the future at once.

`NeuralProphet` incorporates conformal prediction techniques into its forecasting workflow, specifically the **inductive (split) conformal prediction (ICP)** approach. The key steps are as follows:

1. **Training set**: This is the initial dataset that's used to train the forecasting model. `NeuralProphet` uses this data to create an initial PI.

2. **Calibration set**: This set refines the PIs established from the training set. After training the model, `NeuralProphet` evaluates the accuracy of its predictions by comparing the actual target variable values with the predicted outputs.

3. **Quantifying uncertainty**: The difference between actual and predicted values helps `NeuralProphet` measure the uncertainty in its forecasts. This is a crucial step as understanding this variance is essential for generating precise PIs.

4. **Final PI formation**: After quantifying the uncertainty using the calibration set, `NeuralProphet` formulates the final PI. This interval provides a range within which the actual future values are expected to lie with a predefined confidence level.

`NeuralProphet` integrates conformal prediction methods into its forecasting workflow, particularly employing the ICP strategy. This approach enables the creation of statistically robust uncertainty sets or intervals for model predictions, enhancing their reliability and confidence.

Let's look at two methodologies that are utilized by `NeuralProphet` within the conformal prediction framework to establish PIs:

- **Quantile regression**: This technique enables the algorithm to learn only specific quantiles of output variables for each instance. By default, `NeuralProphet` provides a singular output as a point estimate for each instance, calculated based on a 50th-percentile regression. The `NeuralProphet` object requires at least one upper and lower quantile pair as parameters to create a PI. For instance, for a 90% probability of the actual value falling within the estimated interval, the confidence level is set at 0.9, defining two quantiles at 0.05 and 0.95, corresponding to the 5th and 95th percentiles of the forecast distribution.

- **Conformal prediction**: This method introduces a calibration process to the existing model to ascertain uncertainties in point estimators and PIs. Post creation and splitting the data into training and calibration sets for a `NeuralProphet` model, the `conformal_predict` method can be utilized to produce a conformal forecast. `NeuralProphet` uses two variants of UQ – naïve conformal prediction and **conformalized quantile regression (CQR)**, which we looked at in *Chapter 7*.

To demonstrate how `NeuralProphet` can create PIs using conformal prediction, we will follow the notebook at `https://github.com/PacktPublishing/Practical-Guide-to-Applied-Conformal-Prediction/blob/main/Chapter_08_NeuralProphet.ipynb`, which is based on UQ in the `NeuralProphet` tutorial (`https://neuralprophet.com/how-to-guides/feature guides/uncertainty_quantification.html`).

The dataset uses hospital electric load data from a San Francisco hospital electric load dataset (`https://github.com/ourownstory/neuralprophet-data`).

Let's look at the header of the dataset:

	ds	y
	2015-01-01 01:00:00	778.007969
	2015-01-01 02:00:00	776.241750
	2015-01-01 03:00:00	779.357338
	2015-01-01 04:00:00	778.737196
	2015-01-01 05:00:00	787.835835

Figure 8.6 – San Francisco hospital load dataset

`NeuralProphet` requires the data in the specific format with a time column named *ds* and time series values in a column called *y*.

Let's create a `NeuralProphet` object that specifies a data splitting ratio between 0 and 1:

```
m = NeuralProphet()
train_df, test_df = m.split_df(df, freq="H", valid_p=1.0 / 16)
```

By default, `NeuralProphet`'s forecasting provides a singular output: a point estimate for each instance. This estimate is derived from a 50th percentile regression. A `NeuralProphet` object requires at least one upper and lower quantile pair as its parameters to establish a PI. Yet, within a `NeuralProphet` model, we can define multiple quantiles as desired.

For example, say we want to forecast a hospital's electric load with 90% PIs. We want 90% of the actual values to fall within the generated intervals.

We could train a quantile regression model to predict three quantiles – the 5th, 50th, and 95th percentiles. The 5th and 95th quantiles would provide the lower and upper bounds for 90% PIs. The 50th quantile would provide the median point forecast:

```
confidence_lv = 0.9
quantile_list = [round(((1 - confidence_lv) / 2), 2),
round(((confidence_lv + (1 - confidence_lv) / 2), 2)]
qr_model = NeuralProphet(quantiles=quantile_list)
qr_model.set_plotting_backend("plotly-static")
```

Quantile regression is used in `NeuralProphet` to generate PIs. It trains the model using a specialized loss function called pinball loss, also known as quantile loss.

Unlike simple error minimization, pinball loss weights errors asymmetrically based on the quantile. Under-prediction is penalized more heavily than over-prediction for an upper quantile such as 90%. The opposite is valid for a lower quantile, such as 10%.

This matches the inherent meaning of the quantile – 90% indicates we expect 90% of the actual values to lie below the prediction. So, errors where the actual value exceeds the forecast violate that more significantly.

By minimizing the asymmetric pinball loss during training, the model learns quantile lines that reflect the appropriate probability of actual values falling above or below based on the data. The upper and lower quantiles then form the PI.

We can now fit the model and create a DataFrame with the results, forecasting 30 periods:

```
metrics = qr_model.fit(df, freq="H")
future = qr_model.make_future_dataframe(df, periods=30, n_historic_
predictions=100)
forecast = qr_model.predict(df=future)
```

We can visualize the PIs from quantile regression using a plot. The solid line shows the median forecast, while the shaded area depicts the interval between the lower and upper quantile lines. This represents the range expected to contain the actual values with the specified confidence level:

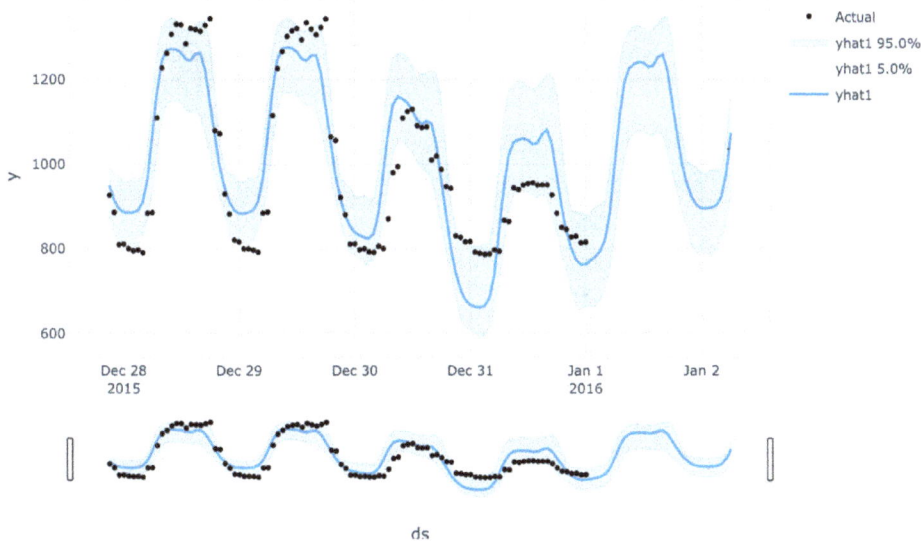

Figure 8.7 – Forecasting hospital electric load using quantile regression

In summary, quantile regression allows NeuralProphet to generate PIs by training the model to forecast quantiles that serve as interval bounds. The pinball loss function enables quantile-based UQ.

Quantile regression relies on modeling assumptions and requires specifying the quantiles of interest. Next, we'll explore how NeuralProphet can produce distribution-free PIs using conformal prediction techniques.

Conformal prediction in NeuralProphet

NeuralProphet employs the split conformal prediction method. This approach necessitates a holdout or calibration set. The dataset must be divided into three sets to execute split conformal prediction: training, calibration, and testing. An initial PI is established using the model trained on the training dataset. Uncertainty is then gauged by comparing the calibration set's target variables with the predicted values. This quantified uncertainty is subsequently incorporated into both ends of the prediction value to form the final conformal PI.

Within NeuralProphet, you can choose either the naïve (based on the absolute residual) or CQR for conformal prediction.

Add a calibration set using the data splitting function:

```
train_df, cal_df = m.split_df(train_df, freq="H", valid_p=1.0 / 11)
```

You can build any NeuralProphet model you deem fit as the base model. The calibration process in conformal prediction would be later added to the base model to quantify the uncertainty in our final estimation. We are interested in knowing how conformal prediction affects different models.

In our example, we will compare the conformal prediction results between a simple quantile regression and a complex four-layer autoregression model in our illustration.

We will specify 72 hours as lags and create a simple quantile regression model as base model 1. We will also create a four-layer autoregression model as base model 2:

```
n_lags = 3 * 24
cp_model1 = NeuralProphet(quantiles=quantile_list)
cp_model1.set_plotting_backend("plotly-static")
cp_model2 = NeuralProphet(
    yearly_seasonality=False,
    weekly_seasonality=False,
    daily_seasonality=False,
    n_lags=n_lags,
    ar_layers=[32, 32, 32, 32],
    learning_rate=0.003,
    quantiles=quantile_list,
)
cp_model2.set_plotting_backend("plotly-static")
```

After configuring the model, we must fit the model with the train set. Suppose you have further split the training dataset into training and validation. In that case, you can either concatenate the two datasets in one dataset for training or assign the training and validation datasets as two separate parameters.

Feed the training subset in the configured NeuralProphet models. Then, configure the hourly frequency by assigning H to the freq parameter:

```
set_random_seed(0)
metrics1 = cp_model1.fit(train_df, freq="H")
set_random_seed(0)
metrics2 = cp_model2.fit(train_df, freq="H")
```

Let's use the fitted base model to forecast both the point prediction and the quantile regression PIs for the testing dataset:

```
forecast1 = cp_model1.predict(test_df)[n_lags:]
forecast2 = cp_model2.predict(test_df)[n_lags:]
```

Option 1 – naïve conformal prediction

After training the base model, we can carry out the calibration process using the naïve module. The steps are as follows:

1. Predict the output value of the instances within the calibration set.

2. Calculate the absolute residual by comparing the actual and predicted values for each observation in the calibration set.

3. Sort all residuals in ascending order.

4. Find the quantile of the distribution of the absolute residuals with the desired confidence level.

5. Use the quantile of the distribution of the absolute residuals to make the final PIs.

Returning to our example, we need to denote the parameter value for the calibration set and the significant level (alpha) for conformal prediction on top of our pre-trained models:

```
Method = ""aïve"
alpha-= 1 - confidence_lv
```

We can now enable conformal prediction using pre-trained models.

```
naïve_forecast1 = cp_model1.conformal_predict(
    test_df,
    calibration_df=cal_df,
    alpha=alpha,
    method=method,
    plotting_backend="plotly-static",
    show_all_PI=naïvee,
)
naive_forecast2 = cp_model2.conformal_predict(
    test_df,
    calibration_df=cal_df,
    alpha=alpha,
    method=method,
    plotting_backend""plotly-static",
    show_all_PI=True,
)
```

NeuralProphet can plot the one-sided interval width versus the selected confidence level:

Naive One-Sided Interval Width with q

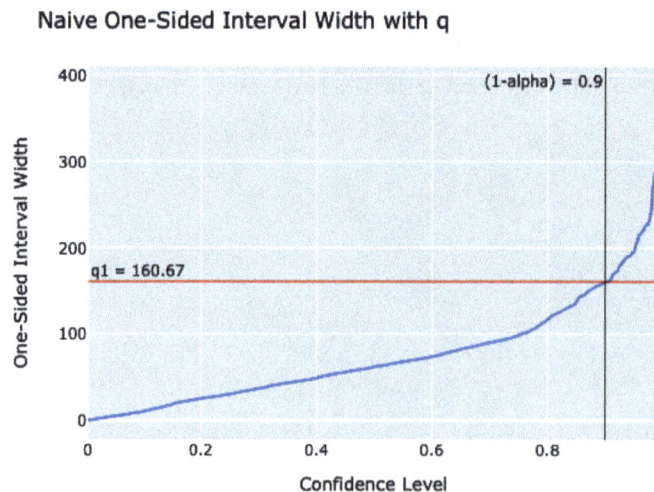

Figure 8.8 – The one-side interval width versus the confidence level

This plot demonstrates how the width of the PI changes with different confidence levels (1-alpha).

Option 2 – CQR

CQR operates in the following manner within the CQR module:

1. Non-conformity scores are computed as the disparities between data points from the calibration dataset and their closest prediction quantile. These scores offer insight into the fit of the data to the existing quantile regression model. Data points within the quantile regression interval yield negative non-conformity scores, while those outside the interval produce positive scores.

2. The non-conformity scores are then organized in order.

3. The alpha value is determined so that a fraction of the scores greater than alpha matches the error rate.

4. An amount of alpha adjusts the regression model's quantiles.

Based on alpha's value, the CQR model can be interpreted in two ways.

When the one-sided PI width adjustment is positive, the CQR expands beyond the QR intervals. This suggests that the CQR perceives the QR interval as overly confident.

On the other hand, if the adjustment is negative, the CQR narrows the QR intervals, indicating that the QR interval might be overly cautious.

We can run the CQR option using the following code:

```
method = "cqr"
cqr_forecast1 = cp_model1.conformal_predict(
    test_df, calibration_df=cal_df, alpha=alpha, method=method,
plotting_backend="plotly-static"
)
cqr_forecast2 = cp_model2.conformal_predict(
    test_df, calibration_df=cal_df, alpha=alpha, method=method,
plotting_backend="plotly-static"
)
```

Again, we can plot the PIs to examine how this CQR method affects the result:

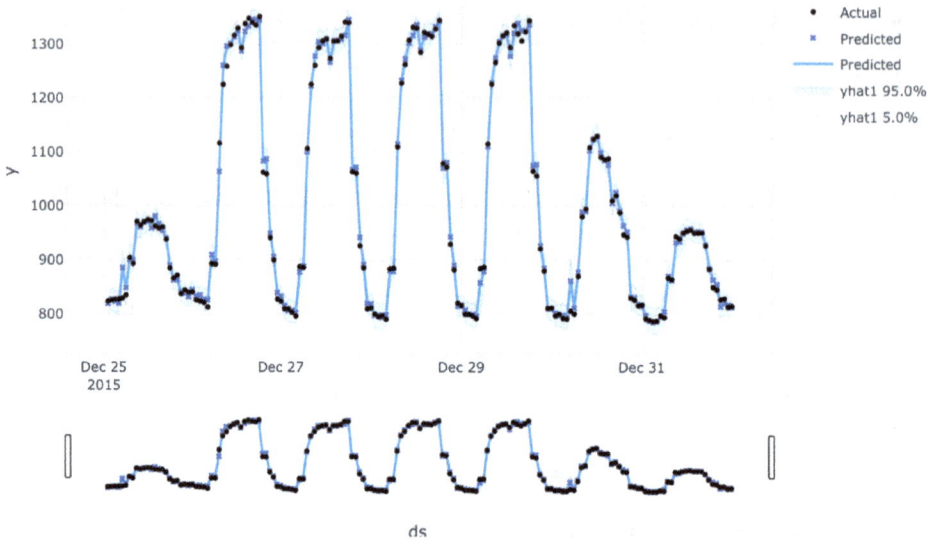

Figure 8.9 – Forecasting hospital electric load using quantile regression (the CQR option)

We can see that `NeuralProphet` has produced excellent PIs using conformal prediction with the `cqr` option.

Now, let's learn how to evaluate performance and obtain some insights by comparing various UQ methods we have used.

We are using the interval width and miscoverage rate as the performance metrics:

- `interval_width`: This is the average PI, or `q_hat`, multiplied by two because it is static or non-adaptive; this is also known as the **efficiency metric**

- `miscoverage_rate`: This is the actual miscoverage error rate on the OOS test set; this is also known as the **validity metric**

Let's evaluate the models we trained earlier. Based on the results in the notebook, we obtain the following conclusions:

- Quantile regression does not provide the required coverage

- The more complex the underlying model, the more accurate it is, hence the lower *interval width* of CP intervals

- For the default model, CQR outputs a narrower *PI width* than naïve

Conformal prediction with Nixtla

We will use the notebook at `https://github.com/PacktPublishing/Practical-Guide-to-Applied-Conformal-Prediction/blob/main/Chapter_08_NixtlaStatsforecast.ipynb` to illustrate how to use conformal prediction to create PIs for popular statistic and econometrics models.

We will use the hourly dataset from the M4 Competition:

1. First, let's install Nixtla's `statsforecast`:

   ```
   !pip install git+https://github.com/Nixtla/statsforecast.git
   ```

2. Then, we must import the necessary modules, including specifically Nixtla's modules:

   ```
   from statsforecast.models import SeasonalExponentialSmoothing,
   ADIDA, and ARIMA
   from statsforecast.utils import ConformalIntervals
   import matplotlib.pyplot as plt
   from statsforecast.models import (
       AutoETS,
       HistoricAverage,
       Naive,
       RandomWalkWithDrift,
       SeasonalNaive
   )
   ```

3. Next, we must load the train and test datasets:

   ```
   train = pd.read_csv('https://auto-arima-results.s3.amazonaws.
   com/M4-Hourly.csv')
   test = pd.read_csv('https://auto-arima-results.s3.amazonaws.com/
   M4-Hourly-test.csv'
   ```

4. Let's look at the dataset's structure. Similar to `NeuralProphet`, `statsforecast` requires columns to be named in a specific way:

	unique_id	ds	y
0	H1	1	605.0
1	H1	2	586.0
2	H1	3	586.0
3	H1	4	559.0
4	H1	5	511.0

Figure 8.10 – Hourly dataset from the M4 Competition

5. We can now train the models; we will only use the first eight series of the dataset to reduce the total computational time:

```
n_series = 8
uids = train['unique_id'].unique()[:n_series]
train = train.query('unique_id in @uids')test = test.
query('unique_id in @uids')
```

Now, we can plot the series:

```
StatsForecast.plot(train, test, plot_random = False)
```

We will get the following output:

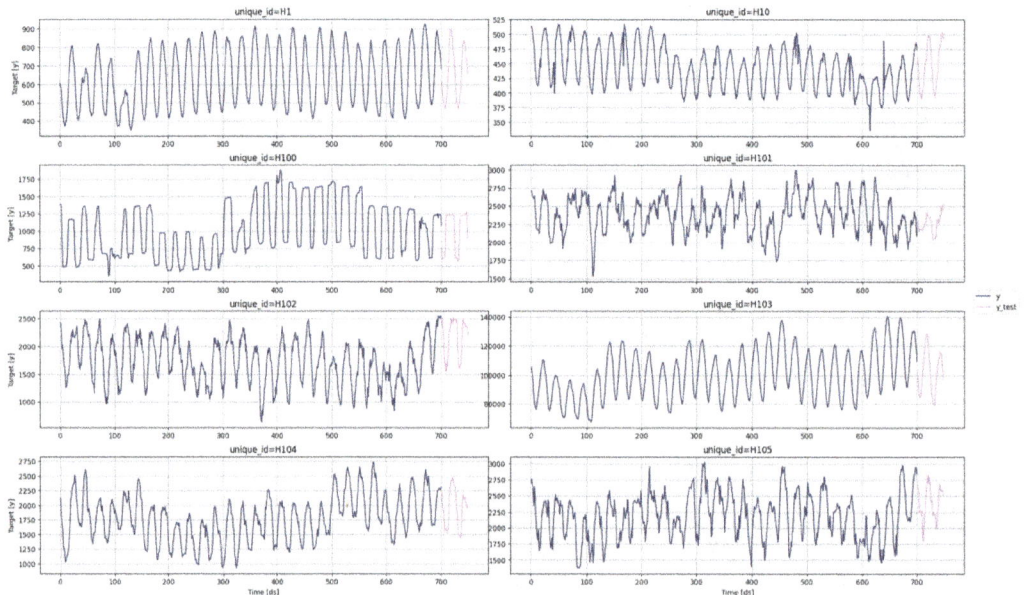

Figure 8.11 – Hourly series from the M4 Competition

6. Let's create a list of models and instantiation parameters. To use these models, we need to import them from `statsforecast.models` and then instantiate them. Given that we're working with hourly data, we need to set `seasonal_length=24` in the models that require this parameter:

```
models = [
    AutoETS(season_length=24),
    HistoricAverage(),
    Naive(),
    RandomWalkWithDrift(),
    SeasonalNaive(season_length=24)
]
```

To instantiate a new `StatsForecast` object, we need the following parameters:

- `df`: The DataFrame that contains the training data.

- `models`: The list of models defined in the previous step.

- `freq`: A string indicating the frequency of the data. See pandas' available frequencies.

- `n_jobs`: An integer that indicates the number of jobs used in parallel processing. Use `-1` to select all cores:

```
sf = StatsForecast(
    df=train,
    models=models,
    freq='H',
    n_jobs=-1
)
```

7. Now, we're ready to generate the point forecasts and the PIs. To do this, we'll use the forecast method, which takes two arguments:

- `h`: An integer that represents the forecasting horizon. In this case, we'll forecast the next 48 hours.

- `level`: A list of floats with the confidence levels of the PIs. For example, `level= [95]` means that the range of values should include the actual future value with a probability of 95%:

```
levels = [80, 90, 95, 99] # confidence levels of the prediction
intervals
forecasts = sf.forecast(h=48, level=levels)
forecasts = forecasts.reset_index()
forecasts.head()
```

8. Now, we can plot the PIs:

```
sf.plot(train, test, plot_random = False,
models=['SeasonalNaive'], level=levels)
```

Here's the plot for this:

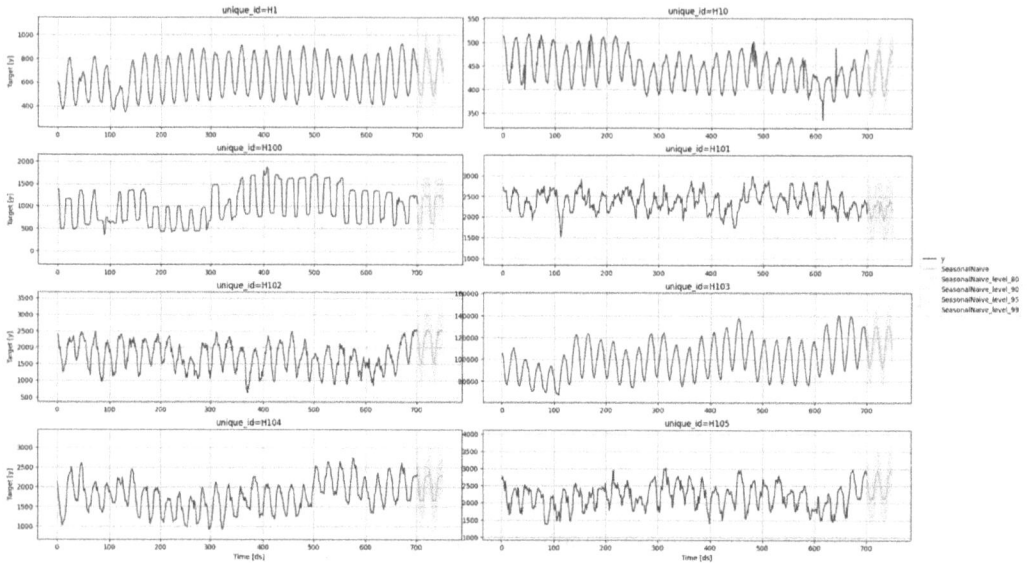

Figure 8.12 – Forecasting hourly series with a seasonal naïve benchmark

Multi-quantile losses and statistical models can provide PIs. Still, the problem is that these are uncalibrated, meaning that the actual frequency of observations falling within the interval does not align with its confidence level. For example, a calibrated 95% PI should contain the actual value in repeated sampling 95% of the time. On the other hand, an uncalibrated 95% PI might contain the true value only 80% of the time or 99% of the time. In the first case, the interval is too narrow and underestimates the uncertainty, while in the second case, it is too broad and overestimates the uncertainty.

Statistical methods also often assume normality. Here, we calibrated PIs produced by statistical models using conformal prediction. Conformal PIs use cross-validation on a point forecaster model to generate the intervals. No prior probabilities are needed, and the output is well calibrated. No additional training is required, and the model is treated as a black box. The approach is compatible with any model. `Statsforecast` now supports conformal prediction on all available models.

`StatsForecast` can train multiple models on different time series efficiently. These models can generate probabilistic forecasts, producing point forecasts and PIs. For this example, we'll use `SimpleExponentialSmoothing` and ADIDA (a model for intermittent demand), which do not provide a PI natively. Thus, using conformal prediction to generate the PI makes sense. We'll also show how to use it with ARIMA to provide PIs that don't assume normality.

To use these models, we first need to import them from `statsforecast.models`, after which we need to instantiate them, as follows:

```
intervals = ConformalIntervals(h=24, n_windows=2)
models = [
    SeasonalExponentialSmoothing(season_length=24,alpha=0.1,
prediction_intervals=intervals),
    ADIDA(prediction_intervals=intervals),
    ARIMA(order=(24,0,12), season_length=24, prediction_
intervals=intervals),
]
sf = StatsForecast(
    df=train,
    models=models,
    freq='H',
)
levels = [80, 90] # confidence levels of the prediction intervals
forecasts = sf.forecast(h=24, level=levels)
```

Let's plot the forecasts that are produced using conformal prediction for ARIMA:

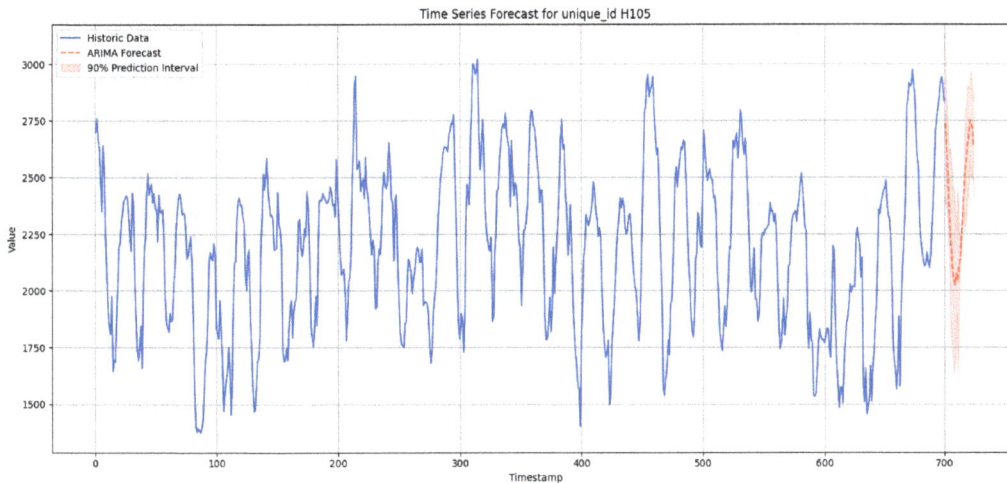

Figure 8.13 – PIs produced for ARIMA using conformal prediction

This section explored implementations of conformal prediction for time series forecasting in several popular open source libraries.

Amazon Fortuna provides conformal prediction capabilities through its EnbPI module. This allows us to generate non-parametric PIs by wrapping any ensemble model with bootstrap resampling. We saw how EnbPI leverages an ensemble to approximate the forecast distribution without assumptions.

Nixtla, an open source library for time series modeling, includes conformal prediction functions for forecasting tasks. We examined how its CP module can take any underlying model and add conformal PIs. Nixtla also supports online conformal prediction for adaptive intervals.

Finally, `NeuralProphet` natively integrates conformal prediction and quantile regression to quantify uncertainty. We examined its ICP approach, which uses a calibration set to refine initial intervals. This generates valid prediction regions without relying on distributional assumptions.

By incorporating conformal prediction, these libraries make robust and accessible UQ available to time series forecasters in Python. The diversity of implementations demonstrates the flexibility of conformal prediction as a model-agnostic framework that can be applied to any forecasting method.

Summary

This chapter taught you how to apply conformal prediction to time series forecasting. Conformal prediction is a powerful technique for crafting PIs for point forecasting models.

This chapter also offered insights into how to harness this method using open source platforms.

We began by exploring UQ in a time series, delving into the significance of PIs, and showcasing various strategies to generate them.

The concept of conformal prediction and its application in forecasting scenarios was central to this chapter. At this point, you are equipped with the knowledge to apply these methodologies in real-world settings, empowering your forecasting models with precise uncertainty bounds. Adding confidence measures to predictions ensures that the forecasts are accurate and reliable.

With a solid understanding of conformal prediction for time series, we will now focus on another critical application area – computer vision.

9

Conformal Prediction for Computer Vision

In today's fast-paced world, computer vision has grown beyond mere image recognition to be a fundamental cornerstone in numerous real-world applications. From self-driving cars navigating bustling streets to medical imaging systems that detect early signs of diseases, the demand for reliable and accurate computer vision models has never been higher. However, with the increasing complexity of these systems and their applications, a critical need arises for the ability to quantify the uncertainty associated with their predictions.

Enter **conformal prediction**, a ground-breaking framework that offers a robust means to encapsulate the uncertainty inherent in machine learning models. While traditional computer vision models often produce a singular prediction, the true power of conformal prediction lies in its ability to provide a set of possible outcomes, each backed by a confidence level. This offers practitioners a more informed, nuanced view of the model's predictions, enabling safer and more reliable deployments in critical applications.

This chapter dives deep into the marriage of conformal prediction and computer vision. We begin by illuminating the necessity of uncertainty quantification in computer vision, highlighting its significance in real-world scenarios including **autonomous driving** and **healthcare diagnostics**. As we navigate deeper, we'll explore the Achilles' heel of modern deep learning models: *their tendency to produce miscalibrated predictions*.

By the journey's end, you'll gain hands-on experience building state-of-the-art computer vision classifiers imbued with the power of conformal prediction. We'll introduce and guide you through the best open source conformal prediction libraries in computer vision applications to ensure you have all the tools necessary to embark on this journey.

In this chapter, we're going to cover the following main topics:

- Uncertainty quantification for computer vision
- Why deep learning produces miscalibrated predictions

- Various approaches to quantify uncertainty in computer vision problems
- Conformal prediction for computer vision
- Building computer vision classifiers using Conformal prediction

Uncertainty quantification for computer vision

As a domain, computer vision has transformed many sectors by automating complex tasks that were once reserved for human eyes and cognition. Computer vision models have become an integral part of modern technology, whether it's detecting pedestrians on the road, identifying potential tumours in medical scans, or even analyzing satellite images for environmental studies. However, as the reliance on these models grows, so does the need to understand and quantify the uncertainty associated with their predictions.

Why does uncertainty matter?

Before deep-diving into the mechanics, it's essential to understand why we need **uncertainty quantification (UQ)** in the first place. Let's go through some of the reasons as follows:

- **Safety and reliability**: A wrong prediction can have severe consequences in critical applications, such as medical imaging or autonomous driving. Knowing the confidence level in a prediction can aid in decision-making, such as whether to trust the model's prediction or seek human intervention.

- **Model improvements**: Uncertainty measurements can provide insights into areas where the model might be lacking, helping to guide data collection and training enhancements.

- **Trustworthiness**: Knowing that a system acknowledges its limitations and can provide confidence intervals or uncertainty metrics for end users and stakeholders makes it more trustworthy.

Navigating the world of computer vision, one inevitably encounters uncertainties that can influence the accuracy of model predictions. But what are the sources of these uncertainties, and can they be managed? Let's delve into the two primary types of uncertainty in computer vision.

Types of uncertainty in computer vision

Uncertainty in computer vision can be broadly classified into two categories:

- **Aleatoric uncertainty**: This type of uncertainty arises from the inherent noise in the data. For instance, low-light images, blurry images, or images taken from varying angles introduce variability that the model might find challenging to handle. Aleatoric uncertainty is often irreducible, meaning no matter how good the model becomes, this uncertainty will always exist due to the inherent noise in the observations.

- **Epistemic uncertainty**: This type of uncertainty stems from the model itself. It could be due to incomplete training data, model architecture choices, or the optimization process. Given enough data or improvements in model design, epistemic uncertainty can be reduced.

In the realm of computer vision, it's not enough to simply get a prediction. As advanced as our models are, they can sometimes be overly confident, potentially leading to misinformed decisions. How do we gauge the reliability of these predictions? Enter the world of uncertainty quantification.

Quantifying uncertainty

Modern computer vision models, especially deep learning architectures, produce predictions that are often overconfident. This miscalibration can be misleading, especially in critical applications. The need, therefore, is not just to produce a prediction but also to accompany it with a measure of confidence or uncertainty.

Various methods have been proposed to quantify uncertainty, ranging from Bayesian neural networks, which provide a distribution over model parameters, to ensemble methods, which rely on the variability of predictions across different models.

However, as we'll see in the subsequent sections, Conformal prediction offers a fresh and rigorous perspective on uncertainty quantification tailored to the needs of computer vision applications.

Uncertainty quantification for computer vision is not a theoretical exercise but a crucial aspect of building reliable, safe, and trustworthy models. Understanding and accounting for their inherent uncertainties will be paramount as computer vision systems continue to permeate every sector.

Why does deep learning produce miscalibrated predictions?

The **ImageNet Large Scale Visual Recognition Challenge (ILSVRC)** is an annual competition where research teams evaluate their algorithms on a given dataset, aiming to push the boundaries of computer vision. 2012 was a watershed moment for the field, marking a significant shift towards the dominance of deep learning in computer vision (https://www.image-net.org/challenges/LSVRC/2012/).

Before the advent of deep learning, computer vision primarily relied on hand-engineered features and traditional machine learning techniques. Algorithms such as **Scale-Invariant Feature Transform (SIFT)**, **Histogram of Oriented Gradients (HOG)**, and **Speeded-Up Robust Features (SURF)** were commonly used to extract features from images. These features would then be fed into machine learning classifiers such as **Support Vector Machines (SVM)** to make predictions. While these methods had their successes, they had significant limitations regarding scalability and performance on more complex datasets.

In 2012, a deep convolutional neural network named AlexNet (https://en.wikipedia.org/wiki/AlexNet), developed by Alex Krizhevsky, Ilya Sutskever, and Geoffrey Hinton, was entered into the ILSVRC. It achieved a top-5 error rate of 15.3%, a staggering 10.8 percentage points lower than the second-place finisher. This dramatic improvement was an incremental step and a quantum leap in performance.

Why was AlexNet revolutionary?

- **Deep architecture**: AlexNet was considerably deeper than other networks of its time. It had five convolutional layers followed by three fully connected layers. This depth allowed it to learn more complex and hierarchical features from the ImageNet dataset.

- **GPU training**: The team utilized **graphics processing units** (**GPUs**) to train the network, which made it feasible to process the massive amount of data in the ImageNet dataset and efficiently train the deep architecture.

- **ReLU activation**: Instead of traditional tanh or sigmoid activation functions, AlexNet employed the **Rectified Linear Unit** (**ReLU**) activation. This choice helped combat the vanishing gradient problem, enabling the training of deeper networks.

- **Dropout**: To prevent overfitting, AlexNet introduced the dropout technique, where random subsets of neurons were "dropped out" during training, forcing the network to learn redundant representations.

The dawn of 2012 marked a transformative moment in the realm of computer vision. Propelled by the unprecedented achievements of AlexNet in the ImageNet competition, the entire industry pivoted towards deep learning, especially **convolutional neural networks** (**CNNs**). As we journey through the aftermath of this revolution, we'll witness the exponential growth in research, widespread industry adoption, and the relentless quest for more data and computational power.

Post-2012 – the deep learning surge

The 2012 ImageNet competition, marked by the triumph of AlexNet, became a watershed moment in the field of computer vision. This victory underscored the profound potential of deep learning, especially **convolutional neural networks** (**CNNs**). As a result, the following happened:

- **Research boom**: After 2012, there was an explosion of research into deep learning for computer vision. Variants and improvements upon AlexNet, such as VGG, GoogLeNet, and ResNet, were rapidly developed, pushing the envelope further.

- **Industry adoption**: Tech giants and start-ups began investing heavily in deep learning research and applications, from facial recognition systems to augmented reality.

- **Datasets and compute**: The success of deep learning fueled the creation of even larger datasets and a race for more powerful computation infrastructure, further accelerating the innovation cycle.

The 2012 ImageNet competition was a turning point, heralding the era of deep learning in computer vision. The principles and breakthroughs of AlexNet laid the foundation for the subsequent advancements we see today, from self-driving cars to real-time video analytics.

The "calibration crisis" in deep learning – a turning point in 2017

Since its triumphant ascendance following the 2012 ImageNet competition, deep learning experienced rapid advancements and widespread adoption across many domains. The community was engrossed in developing architectures, optimization techniques, and applications for five consecutive years. Yet, amidst this whirlwind of innovation, a significant concern remained largely overlooked: *the miscalibration of predictions produced by deep learning systems.*

In real-world applications where automated systems drive decisions, it's not enough for classification networks to merely provide accurate results. These systems play an integral role in various critical sectors, from healthcare to finance, and a misjudgment can have significant consequences. Therefore, it's crucial that these classification networks not only deliver precise outcomes but also possess the self-awareness to flag potential uncertainties or errors in their predictions.

For instance, in a medical diagnostic tool, beyond correctly identifying a disease, the system should also indicate its confidence level in that diagnosis. Medical professionals can take appropriate precautions if uncertain, perhaps seeking additional tests or expert opinions.

Take another example: a self-driving car equipped with a neural network designed to identify pedestrians and various obstacles on the road. In such a scenario, the car's system doesn't just need to recognize people or obstructions; it must do so accurately and in real time. Any delay or misidentification could lead to potentially dangerous situations.

Furthermore, it's not only about detecting obstacles but also understanding the level of certainty in that detection. Imagine a scenario where a self-driving car's detection network struggles to determine whether there's an obstruction ahead confidently. If the car's system is uncertain about an object—perhaps due to poor lighting conditions or an obscured view—it should be programmed to lean more heavily on data from its other sensors, such as lidar or radar, to decide whether braking is necessary and to proceed cautiously, slow down, or even halt. This dual requirement of precise detection and self-awareness of its own certainty levels ensures safer navigation and decision-making, especially in dynamic and unpredictable road environments. See the article *Risk-Sensitive Decision-Making for Autonomous-Driving* (`https://uu.diva-portal.org/smash/get/diva2:1698692/FULLTEXT01.pdf`) if you are interested in more details on the subject.

Accurate confidence estimates play a pivotal role in enhancing model interpretability. Humans inherently understand and relate to probabilities, making it an intuitive measure to gauge predictions.

When a model provides well-calibrated confidence levels, it offers an additional layer of information that bolsters its credibility to the user. This is especially crucial for neural networks, as their decision-making processes can be complex and challenging to decipher. Moreover, reliable probability assessments can be integrated into broader probabilistic models, further expanding their utility and application.

This combination of accuracy and introspection ensures that automated decision-making systems are trustworthy and reliable, fostering confidence in their integration into critical applications.

Miscalibration refers to the disparity between a model's stated confidence in its predictions and the actual accuracy of those predictions. For instance, if a model claims 90% confidence for a set of predictions, one expects approximately 90% of those predictions to be correct. However, despite their high accuracy, deep learning models often needed to catch up on their expressed confidence and actual correctness.

Fast forward to the present, and while contemporary neural networks have seen significant advancements in accuracy compared to those from a decade ago, it's intriguing to note that they no longer maintain calibration.

It wasn't until 2017 that the magnitude of this issue was brought to the forefront of the AI community's attention. A pivotal paper, *On Calibration of Modern Neural Networks (Guo, 2017)*, (`https://proceedings.mlr.press/v70/guo17a.html`) discovered that deep neural networks are poorly calibrated, spotlighting the calibration conundrum inherent in deep learning systems.

This research not only underscored the severe miscalibration of these systems but also brought to light a startling revelation: several of the *state-of-the-art* techniques that had been hailed as breakthroughs, such as dropout, weight decay, and batch normalization, were paradoxically exacerbating the miscalibration issue.

This seminal paper served as a wake-up call. It prompted introspection within the community, urging researchers to question and revisit the techniques they had championed. The paper was a critique and an invitation to explore and rectify the issue. Its lucid exposition and profound insights made it a must-read for anyone in the field.

While the years following the 2012 ImageNet competition were characterized by rapid progress and unbridled optimism, the 2017 paper served as a moment of reckoning. It underscored the importance of introspection in science and the continuous need to refine, recalibrate, and, if necessary, rethink our approaches, ensuring that the AI systems we build are accurate and reliably calibrated.

Confidence calibration is the problem of predicting probability estimates that represent the actual outcome. It is crucial for classification models in many applications because good confidence estimates provide valuable information to establish trustworthiness with the user. Good probability estimates can be used for model interpretability, as humans have a natural cognitive intuition for probabilities.

The authors of the paper *On Calibration of Modern Neural Networks* found that increased model capacity and lack of regularization are closely related to the miscalibration phenomenon observed in deep neural networks. Model capacity has increased dramatically over the past few years, with networks having hundreds or thousands of layers and hundreds of convolutional filters per layer. Recent work shows that very deep or wide models can generalize better than smaller ones while exhibiting the capacity to fit the training set easily. However, this increased capacity can lead to overfitting and miscalibration.

Regarding shallow classical neural networks, the paper *On Calibration of Modern Neural Networks* posited that traditional (or shallow) neural networks were well calibrated. This belief stemmed from a highly cited 2005 paper by Niculescu-Mizil and R. Caruana, titled *Predicting good probabilities with supervised learning* (`https://www.cs.cornell.edu/~alexn/papers/calibration.icml05.crc.rev3.pdf`). Presented at the prestigious ICML conference, this paper has amassed over 1,570 citations since its publication. One of the conclusions reached was that shallow (classical) neural networks were "well calibrated."

However, this conclusion about the calibration of shallow neural networks was upended later. In a 2020 study titled *Are Traditional Neural Networks Well-Calibrated?* (`https://ieeexplore.ieee.org/document/8851962`), the authors debunked the widely held belief that shallow neural networks are well-calibrated. Their findings revealed that traditional shallow networks are poorly calibrated, and their ensembles exhibit the same issue. Fortunately, the researchers also highlighted that the calibration of these networks can be significantly enhanced using the Venn-ABERS conformal prediction method we learned about in previous chapters.

Overconfidence in modern deep learning computer vision models

Many deep learning models designed for computer vision predominantly utilize convolution-based architectures. These architectures have propelled the field forward, achieving unprecedented predictive accuracy. However, there's an unintended side effect: *these models often produce overconfident predictions*:

- **Accuracy versus quality**: The relentless pursuit of accuracy in deep learning has led to models that can correctly classify images with remarkable precision. However, accuracy is just one facet of a model's performance. Predictive quality, which encompasses aspects such as the reliability and calibration of predictions, is equally vital.

- **The overconfidence issue**: Even as these models achieve higher accuracy rates, they tend to be excessively confident in their predictions. This means they do so with high confidence when they make an error, indicating they believe strongly in the incorrect prediction.

- **Implications in critical applications**: This overconfidence poses considerable risks, especially in sectors where the stakes are high. Consider healthcare: a misdiagnosis by a computer vision system analyzing medical scans might lead medical professionals to pursue incorrect treatments if made with high confidence. Similarly, an overconfident misinterpretation of a road scene in autonomous vehicles could result in dangerous manoeuvres.

In essence, as the deep learning community pushes the boundaries of accuracy, it's imperative also to address the calibration of these models. Ensuring that they not only make accurate predictions but also gauge the confidence of those predictions appropriately is crucial, especially when these models are employed in life-critical applications.

Various approaches to quantify uncertainty in computer vision problems

Uncertainty quantification in computer vision is crucial for ensuring vision-based systems' reliability and safety, especially when deployed in critical applications. Over the years, various approaches have been developed to address and quantify this uncertainty. Here's a look at some of the most prominent methods:

- **Bayesian Neural Networks (BNNs)**: These neural networks treat weights as probability distributions rather than fixed values. By doing so, they can provide a measure of uncertainty for their predictions. During inference, multiple forward passes are made with different weight samples, producing a distribution of outputs that capture the model's uncertainty.

- **Monte Carlo dropout**: Monte Carlo dropout involves performing dropout during inference. By running the network multiple times with dropout and averaging the results, a distribution over the outputs is obtained, which can be used to gauge uncertainty.

- **Ensemble methods**: Ensemble methods involve training multiple models and aggregating their predictions. The variance in predictions across models can be used as a proxy for uncertainty. This approach is computationally expensive but often leads to more robust uncertainty estimates.

- **Deep Gaussian processes**: Deep Gaussian processes combine deep learning with Gaussian processes to provide a non-parametric way to estimate uncertainty. They offer a rich way to capture complex uncertainties but can be computationally challenging for large datasets.

- **Conformal prediction**: Conformal prediction provides a set of possible outcomes for a prediction, each with a confidence level. This set-based prediction approach is designed to guarantee coverage, meaning that the actual outcome will fall within the predicted set with a probability equal to the confidence level.

- **Calibration techniques**: While not directly measuring uncertainty, calibration techniques such as Platt scaling or temperature scaling ensure that the predicted confidence scores reflect the true likelihood of correctness. A well-calibrated model's predicted probabilities are more interpretable and can be used as a measure of uncertainty.

The superiority of conformal prediction in uncertainty quantification

Quantifying uncertainty is fundamental to building robust and reliable machine learning models. Several methodologies have emerged over the years, each with its own merits. However, conformal prediction stands out as a particularly compelling framework. Let's explain why:

- **Distribution-free framework**: One of the most notable features of conformal prediction is that it doesn't make any assumptions about the distribution of the data. Many uncertainty quantification methods are based on certain probabilistic assumptions or rely on specific data distributions to function effectively. In contrast, conformal prediction remains agnostic to these considerations, making it versatile and widely applicable across diverse datasets.

- **Theoretical guarantees**: conformal prediction offers robust theoretical guarantees for its predictions. Specifically, it provides a set of potential outcomes for a prediction, and each outcome is associated with a confidence level. The framework ensures that the actual outcome will fall within the predicted set with a probability corresponding to the confidence level. This is a powerful assurance, especially in critical applications where understanding the bounds of a prediction is essential.

- **Model independence**: Another significant advantage of conformal prediction is its independence from the underlying model. Whether you're working with a simple linear regression, a complex deep learning architecture, or any other model, conformal prediction can be applied seamlessly. This flexibility ensures that practitioners are open in their choice of model when seeking to quantify uncertainty.

- **Scalability with dataset size**: conformal prediction is not sensitive to the size of the dataset. Whether dealing with a small dataset with limited entries or a massive one with millions of data points, the framework remains effective and reliable. This scalability is especially beneficial in modern applications where data can range from scarce to overwhelmingly abundant.

While numerous approaches exist for uncertainty quantification, conformal prediction emerges as a frontrunner due to its distribution-free nature, robust theoretical underpinnings, model independence, and scalability. For practitioners seeking a robust and reliable method to gauge the uncertainty of their machine learning models, conformal prediction presents a compelling choice.

Conformal prediction for computer vision

In this section, we will dive deeper into the diverse applications of conformal prediction in computer vision. With its broad range of problems, from image classification to object detection, computer vision presents challenges that require precise and reliable machine learning models. As we navigate these applications, we will demonstrate how conformal prediction is a robust tool to quantify the uncertainty associated with these models.

By exploring these practical examples, we aim to underscore the importance of understanding the model's confidence in its predictions. Understanding is crucial, especially when decisions based on these predictions could have significant consequences. Conformal prediction, with its ability to provide a measure of uncertainty, can greatly aid researchers and practitioners in making informed decisions based on the outputs of their models. This improves the system's reliability and paves the way for more transparent and trustworthy AI implementations in computer vision.

Uncertainty sets for image classifiers using conformal prediction

In 2020, researchers from the University of California, Berkeley, published a paper titled *Uncertainty sets for image classifiers using Conformal Prediction* (`https://arxiv.org/abs/2009.14193`).

This was the first time that computer vision researchers applied conformal prediction to the computer vision problem. The paper described the first conformal prediction method explicitly developed for computer vision, RAPS, which is the current state of the art for image classification.

Here are the key points from the paper:

- The paper proposes a new method called **regularized adaptive predictive sets** (**RAPS**) for generating stable prediction sets with neural network classifiers guaranteed to achieve a desired coverage level.

- RAPS modifies an existing conformal prediction algorithm to produce smaller, more stable prediction sets by regularizing the influence of noisy probability estimates for unlikely classes.

- RAPS is evaluated on ImageNet classification using ResNet and other CNN models. It achieves the desired coverage levels while producing prediction sets that are substantially smaller (5 to 10 times smaller) than a standalone Platt scaling baseline.

- The method satisfies theoretical guarantees on coverage and is proven to provide the best performance for selecting fixed-size sets.

- RAPS provides a practical way to obtain prediction sets from any image classifier that can reliably quantify uncertainty and identify complex test examples. The authors suggest applications in areas such as medical imaging and active learning.

Here is a summary of how the RAPS algorithm works:

1. It uses a pre-trained image classifier to compute class probability estimates for images in the calibration set and class probability estimates for a new test image.

2. For every image within the calibration set, RAPS calculates conformity scores, denoted as E_j as follows: $E_j = \sum_{i=1}^{k} \left(\hat{\pi}_{(i)}(x_j) + \lambda 1 \left[i > k_{reg} \right] \right)$. This is achieved by arranging the probability estimates in a descending sequence. The scores are then computed by accumulating these probability estimates, starting from the highest and continuing down to (and including) the probability estimate of the image's actual class. The calculation is illustrated in the following *Figure 9.1*.

3. A high value of λ acts as a deterrent against creating sets that are larger than k_{reg}.

4. As is standard in inductive conformal prediction, the model then computes the 1-alpha quantile of the conformity scores computed on the calibration set.

5. Outputs the k* highest-score classes where the conformity score E_{test} for the test point is greater or equal the 1-alpha quantile.

The following figure illustrates the RAPS method. The figure is from Anastasios N. Angelopoulos' blog *Uncertainty Sets for Image Classifiers using Conformal Prediction*: (`https://people.eecs.berkeley.edu/~angelopoulos/blog/posts/conformal-classification/`).

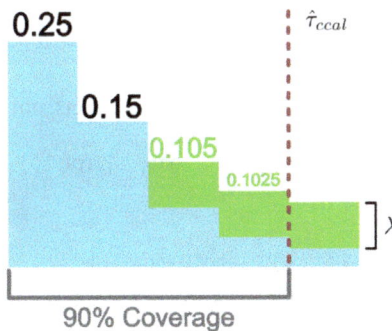

Figure 9.1 – An illustration of the RAPS method (the red line is drawn to achieve exact coverage)

The parameters λ and k_{reg} are estimated by the RAPS model on the calibration set. The intuition behind parameters is that a high λ discourages sets larger than k_{reg}.

By construction, this prediction set provably contains the true class with probability of at least 1-α, where α is the desired error level. The regularization penalty allows RAPS to produce smaller, more stable sets than previous methods such as Platt scaling or the unregularized adaptive method.

This approach allows researchers to use any underlying classifier and produce predictive sets that are assured to meet a designated error rate, such as 90%, all while maintaining a minimal average size. Its ease of deployment makes it a compelling, automated method to gauge the uncertainty of image classifiers, which is crucial in areas including medical diagnostics, autonomous vehicles, and screening hazardous online content.

In summary, RAPS leverages conformal prediction ideas to guarantee coverage, modifies the conformal score to enable smaller sets, and calibrates the procedure correctly using held-out data.

Building computer vision classifiers using conformal prediction

Let's illustrate the application of conformal prediction to computer vision in practice. We will use a notebook from the book repository available at `https://github.com/PacktPublishing/Practical-Guide-to-Applied-Conformal-Prediction/blob/main/Chapter_09.ipynb`. This notebook extensively uses notebooks from Anastasios Angelopolous' *Conformal Prediction* repo at `https://github.com/aangelopoulos/conformal-prediction`.

After loading the data, set up the problem and define the desired coverage and the number of points in the calibration set:

```
n_cal = 1000
alpha = 0.1
```

The softmax scores were split into the calibration and test datasets, obtaining calibration and test labels:

```
idx = np.array([1] * n_cal + [0] * (smx.shape[0]-n_cal)) > 0
np.random.seed(42)
np.random.shuffle(idx)
cal_smx, test_smx = smx[idx,:], smx[~idx,:]
cal_labels, test_labels = labels[idx], labels[~idx]
```

The test dataset contains 49,000 points, and the calibration dataset contains 1,000 points. Both datasets include images and human-readable labels from the ImageNet dataset.

Naïve Conformal prediction

We'll first look at a naïve way to produce prediction sets using conformal prediction:

- Compute a non-conformity score for each calibration point
- Then, an empirical quantile of the calibration scores will be evaluated

This closely resembles what we observed with inductive conformal prediction in earlier chapters. We determine non-conformity scores through hinge loss, and then use the distribution of these scores to calculate the quantile based on the desired coverage. This process, including the final sample correction formula, parallels our approach for inductive conformal prediction:

```
cal_scores = 1-cal_smx[np.arange(n_cal),cal_labels]
q_level = np.ceil((n_cal+1)*(1-alpha))/n_cal
qhat = np.quantile(cal_scores, q_level, method='higher')
```

We can form the prediction sets for test set objects using the computed get adjusted quantile on nonconformity scores:

```
prediction_sets = test_smx >= (1-qhat)
```

The result is an array showcasing sets of predictions. This Boolean array signifies ImageNet classes according to the Boolean values it holds. The Boolean values indicate the classes chosen by the model, with `True` signifying a class is selected and `False` meaning the class is not included in the prediction set.

```
array([[ True, False, False, ..., False, False, False],
       [ True, False, False, ..., False, False, False],
       [ True, False, False, ..., False, False, False],
       ...,
       [False, False, False, ..., False, False,  True],
       [False, False, False, ..., False, False,  True],
       [False, False, False, ..., False, False,  True]])
```

Figure 9.2 – An illustration of the prediction sets for the test set

We can calculate the empirical coverage, which comes very close to the specified confidence level of 90%:

```
empirical_coverage = prediction_sets[np.arange(prediction_sets.
shape[0]),test_labels].mean()
print(f"The empirical coverage is: {empirical_coverage}")
```

We can look at some of the objects and prediction sets.

Figure 9.3 – An object from the test set, the prediction set produced by the
naïve variant of conformal prediction was the label "palace"

For objects with higher levels of uncertainty, prediction sets contain more than one element.

Figure 9.4 – An object from the test set, the prediction set produced by the
naïve variant of conformal prediction was ['Crock Pot', 'digital clock']

The naïve method presents two significant issues:

- Firstly, the probabilities produced by CNNs often need to be more accurate, resulting in sets that don't achieve the intended coverage

- Secondly, for instances where the model lacks confidence, the naive method must include numerous classes to attain the desired confidence threshold, leading to an excessively large set

Temperature scaling isn't a remedy, as it only adjusts the score of the primary class, and calibrating the remaining scores is an overwhelming task. Interestingly, even with the perfect calibration of all scores, the naive approach would still fall short of achieving coverage.

Alternative ways of constructing prediction sets were developed to address these issues, namely **Adaptive Prediction Sets (APS)** and **Regularized Adaptive Prediction Sets (RAPS)**.

Adaptive Prediction Sets (APS)

Next, we'll look at APS, described in the NeurIPS spotlight paper, *Classification with Valid and Adaptive Coverage* (2000) (https://proceedings.neurips.cc/paper/2020/file/244edd7e 85dc81602b7615cd705545f5-Paper.pdf).

In essence, APS presents a simple approach. Instead of directly using the softmax scores, a new threshold is determined based on a calibration dataset. For example, if sets with a projected probability of 93% yield a 90% coverage on the calibration set, then a 93% threshold would be adopted. APS is a particular implementation of RAPS, and unlike the naïve approach, it aims to achieve precise coverage.

However, APS does face a practical hurdle: the average size of its sets is significantly large. Deep learning classifiers grapple with a permutation dilemma: their scores for less certain classes, such as those ranked from 10 to 1,000, don't reflect accurate probability estimates. The arrangement of these classes is largely swayed by noise, prompting APS to opt for vast sets, especially for complex images.

The code describing APS is as follows:

```
# Get scores. calib_X.shape[0] == calib_Y.shape[0] == n
cal_pi = cal_smx.argsort(1)[:, ::-1]
cal_srt = np.take_along_axis(cal_smx, cal_pi, axis=1).cumsum(axis=1)
cal_scores = np.take_along_axis(cal_srt, cal_pi.argsort(axis=1),
axis=1)[
    range(n_cal), cal_labels
]
# Get the score quantile
qhat = np.quantile(
    cal_scores, np.ceil((n_cal + 1) * (1 - alpha)) / n_cal,
method="higher"
)
test_pi = test_smx.argsort(1)[:, ::-1]
test_srt = np.take_along_axis(test_smx, test_pi, axis=1).
cumsum(axis=1)
prediction_sets = np.take_along_axis(test_srt <= qhat, test_
pi.argsort(axis=1), axis=1)
```

Let's look at the code in more detail. It uses APS to generate prediction sets based on a specified quantile threshold:

1. **Calibration phase**:

     I.   `cal_pi = cal_smx.argsort(1)[:, ::-1]`: This sorts the softmax scores `cal_smx` for each instance score from `cal_smx` in descending order and returns the indices of the sorted values.

     II. `cal_srt = np.take_along_axis(cal_smx, cal_pi, axis=1).cumsum(axis=1)`: For each row, it rearranges the scores based on the indices from `cal_pi`, then computes the cumulative sum along the columns.

     III. `cal_scores = np.take_along_axis(cal_srt, cal_pi.argsort(axis=1), axis=1)[range(n_cal), cal_labels]`: This step retrieves the specific scores corresponding to the true labels (`cal_labels`). It first reverts the sorted order of `cal_pi` to get the original ordering and then picks the scores associated with the true labels for each instance.

2. **Determine quantile threshold**:

     I.   `qhat = np.quantile(cal_scores, np.ceil((n_cal + 1) * (1 - alpha)) / n_cal, method="higher")`: Calculates the quantile value based on the provided `alpha`. This value will serve as the threshold for the prediction phase.

3. **Prediction phase**:

I. `test_pi = test_smx.argsort(1)[:, ::-1]`: Similarly, for the test set, it sorts the scores from `test_smx` in descending order and returns the indices of the sorted values.

II. `test_srt= np.take_along_axis(test_smx, test_pi, axis=1).cumsum(axis=1)`: Rearranges the test set scores based on the `test_pi` sorted indices and computes the cumulative sum.

III. `prediction_sets= np.take_along_axis(test_srt <= qhat, test_pi.argsort(axis=1), axis=1)`: For each instance in the test set, it determines which scores are below the quantile threshold qhat. This Boolean array (`test_srt <= qhat`) is then rearranged into its original order using `test_pi.argsort(axis=1)`, resulting in the final prediction sets where `True` entries indicate inclusion in the set.

In essence, this code is used to calibrate model scores to define a threshold and then uses this threshold to generate prediction sets for a new (test) dataset.

We can look at some objects and prediction sets generated by APS.

Figure 9.5 – An object from the test set

Unfortunately, as already mentioned and demonstrated in this example, the prediction sets produced by APS can be vast. The preceding example produced a prediction set of:

['King Charles Spaniel', 'Rhodesian Ridgeback', 'Afghan Hound', 'Basset Hound', 'Bloodhound', 'Redbone Coonhound', 'Otterhound', 'Weimaraner', 'Irish Terrier', 'Norfolk Terrier', 'Norwich Terrier', 'Australian Terrier', 'Dandie Dinmont Terrier', 'Tibetan Terrier', 'Soft-coated Wheaten Terrier', 'Flat-Coated Retriever', 'Golden Retriever', 'Labrador Retriever', 'Vizsla', 'English Setter', 'Irish Setter', 'Gordon Setter', 'Clumber Spaniel', 'English Springer Spaniel', 'Welsh Springer Spaniel', 'Cocker Spaniels', 'Sussex Spaniel', 'Irish Water Spaniel', 'Briard', 'Bullmastiff', 'Leonberger', 'Newfoundland', 'Chow Chow', 'Miniature Poodle', 'Standard Poodle', 'lion', 'brown bear', 'grasshopper', 'leafhopper', 'doormat', 'handkerchief', 'maze', 'prayer rug', 'tennis ball', 'acorn'].

Regularized Adaptive Prediction Sets (RAPS)

We now get hands-on with RAPS, which was briefly introduced in the *Uncertainty sets for image classifiers using conformal prediction* section earlier in this chapter.

We set the RAPS regularization parameters (a larger `lam_reg` value and smaller `k_reg` value leads to smaller sets) and regularization vector in the following code block:

```
lam_reg = 0.01
k_reg = 5
disallow_zero_sets = False
rand = True
reg_vec = np.array(k_reg*[0,] + (smx.shape[1]-k_reg)*[lam_reg,])
[None,:]
```

As previously, we compute non-conformity scores and obtain score quantiles:

```
cal_pi = cal_smx.argsort(1)[:,::-1];
cal_srt = np.take_along_axis(cal_smx,cal_pi,axis=1)
cal_srt_reg = cal_srt + reg_vec
cal_L = np.where(cal_pi == cal_labels[:,None])[1]
cal_scores = cal_srt_reg.cumsum(axis=1)[np.arange(n_cal),cal_L] -
np.random.rand(n_cal)*cal_srt_reg[np.arange(n_cal),cal_L]
qhat = np.quantile(cal_scores, np.ceil((n_cal+1)*(1-alpha))/n_cal,
method='higher')
```

We can deploy predictions on the test set using the following code:

```
n_test = test_smx.shape[0]
test_pi = test_smx.argsort(1)[:,::-1]
test_srt = np.take_along_axis(test_smx,test_pi,axis=1)
test_srt_reg = test_srt + reg_vec
test_srt_reg_cumsum = test_srt_reg.cumsum(axis=1)
indicators = (test_srt_reg.cumsum(axis=1) - np.random.rand(n_
test,1)*test_srt_reg) <= qhat if rand else test_srt_reg.cumsum(axis=1)
-test_srt_reg <= qhat
if disallow_zero_sets: indicators[:,0] = True
prediction_sets = np.take_along_axis(indicators,test_
pi.argsort(axis=1),axis=1)
```

Let's look at some objects and prediction sets generated by RAPS.

Figure 9.6 – An object from the test set; the prediction set produced by RAPS was ['electric ray']

We can see that for objects with little uncertainty, RAPS produces one-element prediction sets. Unlike APS, RAPS still produces rather parsimonious prediction sets for objects involving more uncertainty.

Figure 9.7 – An object from the test set; the prediction set produced by RAPS was ['red wolf', 'coyote', 'dhole', 'gray fox']

Let's summarize the chapter next.

Summary

In the rapidly evolving realm of technology, computer vision has transformed from mere image recognition into an integral component of countless real-world applications. As these applications span diverse fields such as autonomous vehicles and medical diagnostics, the pressure on computer vision models to deliver accurate and reliable predictions intensifies. With the growing sophistication of these models comes a dire need: quantifying prediction uncertainties.

This is where conformal prediction shines. Unlike traditional models that typically output a singular prediction, conformal prediction offers a range of potential outcomes, each coupled with a confidence measure. This novel approach grants users a detailed perspective on model predictions, which is invaluable for applications where precision is paramount.

This chapter delved into the symbiotic relationship between conformal prediction and computer vision. We started by emphasizing the importance of uncertainty quantification in computer vision, citing its pivotal role in areas including autonomous transportation and medical imaging. Further, we shed light on a major area for improvement in contemporary deep learning models: their tendency to deliver miscalibrated predictions.

By working through this chapter, you have acquired the expertise to craft cutting-edge computer vision classifiers infused with the capabilities of conformal prediction. Additionally, you got experience of the top-tier open source conformal prediction tools tailored for computer vision, ensuring you're well equipped for future endeavors.

The key achievements in this chapter are to grasp the role of uncertainty quantification in computer vision, unravel the reasons behind deep learning's miscalibrated predictions, explore diverse strategies to measure uncertainty in computer vision tasks, comprehend the fundamentals and applications of conformal prediction in computer vision, and attain mastery of constructing computer vision classifiers powered by conformal prediction.

In the next chapter, we will navigate the world of conformal prediction in NLP, understand its significance, and learn how to harness its power for more reliable and confident predictions.

10

Conformal Prediction for Natural Language Processing

Natural language processing (**NLP**) grapples with the complexities of human language, where uncertainty is an inherent challenge. As NLP models become integral to risk-sensitive and critical applications, ensuring their reliability is paramount. Conformal prediction emerges as a promising technique, offering a way to quantify the trustworthiness of these models' predictions, particularly when faced with miscalibrated outputs from deep learning models.

In this chapter, we will navigate the NLP conformal prediction world, understand its significance, and learn how to harness its power for more reliable and confident predictions.

In this chapter, we're going to cover the following main topics:

- Uncertainty quantification for NLP
- Why deep learning produces miscalibrated predictions
- Various approaches to quantify uncertainty in NLP problems
- Conformal prediction for NLP
- Building NLP classifiers using conformal prediction
- Open source tools for conformal prediction in NLP

Uncertainty quantification for NLP

Uncertainty quantification in NLP is an essential yet often overlooked aspect of model development and deployment. As NLP models become increasingly integrated into critical applications—from healthcare diagnostics to financial predictions—the need to understand and convey the confidence level of their outputs becomes paramount. Uncertainty quantification provides a framework for assessing the reliability of predictions, allowing users and developers to gauge the model's decisiveness and the potential risks of relying on its results. This section delves into the importance, methodologies, and practical considerations of uncertainty quantification in NLP, highlighting its pivotal role in building robust and trustworthy language models.

We will now explore uncertainty in NLP and the benefits and challenges of quantifying uncertainty in NLP applications.

What is uncertainty in NLP?

NLP, at its core, is about making sense of human language—a medium known for its richness, ambiguity, and diversity. The inherent variability in language usage, context-driven meanings, and the ever-evolving nature of linguistic constructs make NLP tasks inherently uncertain. For instance, the word "bank" could refer to a financial institution or the side of a river depending on the context.

Benefits of quantifying uncertainty in NLP

Quantifying uncertainty in NLP is not just a theoretical exercise; it has the following tangible benefits:

- **Trustworthiness**: Quantifying uncertainty either bolsters confidence in specific predictions or highlights areas of caution.

- **Performance evaluation**: This assesses the efficacy of various models by examining the uncertainty in their metrics.

- **Enhancement opportunities**: It can recognize areas where a model can be refined, especially in contexts such as active learning.

- **Risk management**: By understanding the degree of uncertainty in predictions, stakeholders can make more informed decisions. For instance, an NLP model predicting sentiment might be 80% certain that a review is positive. Knowing this, a business might prioritize addressing reviews where the model's certainty is lower.

- **Model transparency**: A model that can express its uncertainty is perceived as more transparent and trustworthy. Users of the model can better understand when to trust the model's output and when to approach it with caution.

- **Model training**: During the training phase, understanding areas of high uncertainty can guide data collection efforts. If a model is uncertain about a particular data type, gathering more of that data can lead to more robust training.

The challenges of uncertainty in NLP

Despite its importance, managing uncertainty in NLP is challenging. Here are some reasons why:

- **Data sparsity**: Many NLP tasks lack representative data for all possible linguistic variations, leading to models that are uncertain about less common data points

- **Ambiguity in language**: As mentioned, words can have multiple meanings based on context, leading to inherent uncertainty

- **Model complexity**: Advanced models such as deep learning networks can sometimes act as black boxes, making it challenging to discern areas of uncertainty

In building robust NLP systems, understanding and quantifying uncertainty becomes paramount. As we dive deeper into the chapter, we'll explore techniques, particularly **conformal prediction**, that offer a structured approach to tackle these challenges head-on.

Understanding why deep learning produces miscalibrated predictions

In the rapidly evolving field of NLP, deep learning played a pivotal role in enabling machines to process and generate language in ways that were once the exclusive domain of humans. The next section introduces the key concepts and milestones in deep learning that has significantly influenced NLP.

Introduction to deep learning in NLP

Deep learning, a subset of machine learning, relies on neural networks with many layers (hence "deep") to analyze various data factors. In the context of NLP, deep learning has been a game-changer, enabling machines to understand and generate human language with unprecedented accuracy:

- **Evolution of architectures**: The journey began with simpler architectures such as feedforward neural networks and **recurrent neural networks** (**RNNs**). With its ability to remember past information, the latter was particularly influential in sequence-based tasks such as language translation. Later, more advanced architectures such as **long short-term memory** (**LSTM**) and the **Transformer model** further elevated performance standards.

- **BERT and transformers**: The introduction of **Bidirectional Encoder Representations from Transformers** (**BERT**) marked a significant milestone. BERT achieved state-of-the-art results in numerous NLP tasks by analyzing words concerning their entire context (both left and right of a word). The Transformer architecture, which BERT is based on, introduced attention mechanisms that allow models to focus on specific parts of the input text, much like how humans pay attention to particular words when comprehending language.

- **Language models and LLMs: Large language models (LLMs)** such as **generative pre-trained transformer (GPT)** and its iterations, such as ChatGPT, have set new standards in NLP. With billions of parameters, these models can generate human-like text, answer questions, and even assist in creative writing. ChatGPT, in particular, has been influential in creating conversational agents capable of more natural and coherent interactions.

- **Transfer learning and fine-tuning**: One of the revolutionary aspects of these developments is the idea of transfer learning. Models such as BERT and GPT are pre-trained on vast corpora and can be fine-tuned on specific tasks with smaller datasets. This approach has democratized deep learning in NLP, allowing teams with limited resources to achieve competitive results.

With these advancements, deep learning models have become the backbone of many modern NLP applications, from chatbots to search engines. However, as we'll explore in the subsequent sections, their complexity and sheer scale introduce challenges, especially in calibration.

Challenges with deep learning predictions in NLP

Deep learning has undeniably advanced the capabilities of NLP, but it also brings forth several challenges and pitfalls. As we navigate the landscape of deep learning in NLP, we must be aware of these issues. Some of the notable challenges include the following:

- **Model overconfidence**: Deep learning models, given their capacity to fit complex patterns, often become overconfident in their predictions. For instance, in sentiment analysis, a model might predict a text as positive with 60% confidence when, in reality, the actual confidence should be much lower due to ambiguous phrasing.

- **Data distribution shift**: NLP models are often trained on specific datasets and may not be exposed to the full linguistic diversity of real-world inputs. When faced with out-of-distribution data, these models can produce miscalibrated predictions.

- **Lack of explicit uncertainty modeling**: Traditional deep learning approaches don't inherently model uncertainty. They optimize for accuracy, often at the cost of a reliable uncertainty estimate.

- **Complexity and non-linearity**: The intricate architectures of deep learning models, especially with multiple layers and non-linear activations, can sometimes lead to unpredictable behavior, especially when handling edge cases or rare linguistic constructs.

Let's go through the implications of miscalibration next.

The implications of miscalibration

Miscalibration in NLP models is more than just a mere academic concern. In real-world applications, it can lead to misinformed decisions, misplaced trust, and even potentially harmful outcomes, especially in sensitive areas such as healthcare, finance, and legal systems:

- **Decision-making risks**: Overconfident models can lead stakeholders to make decisions based on misguided confidence, potentially causing miscommunications or flawed strategies

- **Loss of trust**: Users might lose faith in an NLP system if it frequently expresses high confidence in incorrect predictions

- **Resource misallocation**: In automated systems, a miscalibrated model might prioritize tasks inefficiently, wasting computational resources on tasks where human intervention would have been more appropriate

Recognizing these challenges is the first step. As we progress, we'll delve into conformal prediction—a technique that presents a viable solution to the miscalibration issues plaguing deep learning models in NLP.

Various approaches to quantify uncertainty in NLP problems

Multiple methods to quantify uncertainty in NLP problems have been explored to address the challenges of miscalibration and language's inherent unpredictability.

We will now look at Bayesian approaches to UQ.

Bayesian approaches to uncertainty quantification

Bayesian methods provide a framework for modeling uncertainty. By treating model parameters as distributions rather than fixed values, Bayesian neural networks offer a measure of uncertainty associated with predictions. This probabilistic approach ensures that the model not only gives an estimate but also conveys the confidence or spread of that estimate.

These are some of the examples of Bayesian approaches to UQ.

- **Variational inference** is a technique to approximate the posterior distribution of the model parameters, enabling the network to output distributions for predictions.

- **Bayesian neural networks** (**BNNs**) are neural networks with weights assigned to probability distributions. By sampling from these distributions, BNNs can produce a range of outputs, reflecting the uncertainty in predictions.

- **Monte Carlo dropout** is a technique wherein dropout is applied during inference. We can gain insight into the model's uncertainty by running the model multiple times and observing the variance in outputs.

Bootstrap methods and ensemble techniques

Bootstrapping involves creating multiple datasets from the original training data through resampling. By training separate models on these datasets, we can capture model uncertainty. This variance across different resamples allows for a more robust evaluation of how changes in the input data can impact predictions.

We will now look at some of the examples of bootstrap methods and model ensembles.

- **Bagging**: Short for bootstrap aggregating, this involves training multiple models on different bootstrap samples. The variance in predictions across models provides an estimate of uncertainty.

- **Model ensembles**: Combining predictions from multiple models can also capture uncertainty. If models trained on the same data but with different architectures disagree on a prediction, it indicates higher uncertainty.

Out-of-distribution (OOD) detection

Identifying inputs that are significantly different from the training data can also help in uncertainty estimation:

- **Likelihood-based methods**: These methods compare the likelihood of new data points to the training data. Lower likelihood indicates higher uncertainty.

- **Adversarial training**: By training models to recognize adversarial examples, we can enhance their ability to identify uncertain inputs.

Understanding and appropriately employing these techniques is crucial in NLP, given human language's inherent ambiguities and nuances. Each approach has its strengths and suitable scenarios, so practitioners must choose wisely based on the specifics of their NLP task.

Conformal prediction for NLP

Conformal prediction is a flexible and statistically robust approach to uncertainty quantification. It is a distribution-free framework that can estimate uncertainty for machine learning models without requiring model retraining or access to limited APIs. The central idea behind conformal prediction is to output a set of predictions containing the correct output with a user-specified probability. Conformal prediction can help quantify the uncertainty associated with the model's predictions in language models.

Conformal prediction is a framework that delivers valid confidence intervals for predictions, irrespective of the underlying machine learning model. In the NLP landscape, with its inherent challenges of ambiguity, context sensitivity, and linguistic diversity, conformal prediction offers a structured way to quantify uncertainty.

Validity and efficiency are the two fundamental principles of conformal prediction. Validity ensures that the prediction regions (or sets) are correct with a predefined probability, while efficiency ensures these regions are as tight as possible.

How conformal prediction works in NLP

The mechanics of conformal prediction are rooted in ordering predictions based on their "strangeness" or non-conformity scores. The idea is to understand how different a new observation is compared to previous ones:

- **Non-conformity score**: This score measures how different a new prediction is from previous predictions for any NLP task. For instance, the non-conformity might be based on the distance from the decision boundary in text classification.

- **P-values**: P-values are calculated based on the non-conformity scores, representing the confidence level of predictions.

Practical applications of conformal prediction in NLP

Conformal prediction isn't just a theoretical construct; its practical applications in NLP are wide-ranging:

- **Sentiment analysis**: When determining the sentiment of a text snippet, conformal prediction can provide a range or set of possible sentiments, each with its confidence level

- **Named entity recognition**: Conformal prediction can give a confidence score on each tagged entity instead of just tagging entities, helping in tasks where precision is critical

- **Machine translation**: Beyond translating text, conformal prediction can offer confidence intervals for different translation choices, aiding in tasks where mistranslations can have significant consequences

Advantages of using conformal prediction in NLP

Conformal prediction, a relatively recent development in uncertainty quantification, brings a fresh perspective and many benefits to NLP. As we venture into an era where the demand for reliable and trustworthy models is ever-increasing, methods such as conformal prediction stand out, promising to address some innate challenges in NLP. Let's delve into the distinct advantages of integrating conformal prediction in NLP tasks:

- **Model agnostic**: One of the strengths of conformal prediction is its compatibility with any machine learning model. Conformal prediction can be applied to any statistical, machine, or deep learning model.

- **Transparent and interpretable**: Conformal prediction doesn't operate as a black box. The non-conformity scores and resulting p-values offer interpretable metrics of uncertainty.

- **Adaptive**: Conformal prediction is adaptive to the data it's applied to. It doesn't make strong distributional assumptions, making it robust despite diverse linguistic data.

The introduction of conformal prediction into the NLP toolkit offers a promising avenue for practitioners to handle the inherent uncertainties of human language. Providing valid and reliable confidence measures helps build more robust and trustworthy NLP systems.

An example of applied conformal prediction on an NLP task, such as IMDB Movie reviews that have been pre-labeled with "positive" and "negative" sentiment class labels based on the review content, has been discussed here: `https://github.com/M-Soundouss/density_based_conformal_prediction/tree/master/imdb`.

Conformal prediction for NLP and LLMs is an emerging and crucial area of research.

A notable contribution to this field is a paper by Kumar et.al., titled *Conformal Prediction with Large Language Models for Multi-Choice Question Answering* (`https://arxiv.org/abs/2305.18404`).

This paper dives deep into how conformal prediction can be instrumental in quantifying uncertainty in language models, thereby paving the way for a more trustworthy and reliable deployment of large language models, especially in scenarios where safety is paramount.

The paper's primary focus is on multiple-choice question-answering tasks. Through a series of experiments, it showcases the efficacy of conformal prediction in deriving uncertainty estimates that are in strong correlation with prediction accuracy.

Diving into the experimental setup, the authors employed the LLaMA-13B model. This model, boasting 13 billion parameters and trained on a staggering 1 trillion tokens, generated predictions for MCQA questions sourced from the `MMLU` benchmark dataset (`https://paperswithcode.com/sota/multi-task-language-understanding-on-mmlu`).

The experiments were structured around a calibration set that trained the conformal prediction model and an evaluation set that tested the model's prowess. A cross-validation approach was adopted to ensure the experiment's integrity, ensuring the calibration and evaluation sets were sampled from a consistent distribution.

The performance metrics were multifaceted, encompassing accuracy, coverage, and efficiency. A pivotal observation was that the softmax outputs from the LLaMA-13B model, while reasonably calibrated on average, exhibited tendencies of underconfidence and overconfidence, particularly at the extremities of the probability distribution. This observation was particularly pronounced in subjects such as formal logic and college chemistry, which inherently possess more ambiguity and complexity, making them challenging for LLMs to navigate accurately.

One of the standout findings was the strong correlation between the uncertainty estimates provided by conformal prediction and prediction accuracy. Such a correlation implies that when the model exhibits higher uncertainty about its predictions, it's more prone to errors. This insight is invaluable for downstream applications such as selective classification. By leveraging these uncertainty estimates, it's feasible to filter out lower-quality predictions, thereby enhancing the overall user experience.

The paper underscores the potential of conformal prediction as a beacon for uncertainty quantification in LLMs. By integrating this approach, LLMs can be more reliable, especially in high-stakes environments, reinforcing their trustworthiness and broadening their applicability.

The second critical paper is *Robots That Ask For Help: Uncertainty Alignment for Large Language Model Planners* (`https://robot-help.github.io`), published by a team of researchers from Princeton University and DeepMind.

In robotics and artificial intelligence, the aspiration to equip robots with the capability to discern when uncertain is a pivotal challenge. The paper addresses this challenge, particularly regarding robots instructed via language. With its inherent flexibility, language offers a natural interface for humans to convey tasks, contextual information, and intentions. It also facilitates humans in providing clarifications to robots when they encounter uncertainties.

Recent advancements have showcased the potential of LLMs in planning. These models can interpret and respond to unstructured language instructions, generating temporally extended plans. The strength of these LLMs lies in their ability to harness the vast knowledge and rich context they have been pre-trained with, leading to enhanced abstract reasoning capabilities. However, a significant impediment with current LLMs is their propensity to "hallucinate." In other words, they tend to generate outputs with high confidence that, while plausible, might be incorrect and not anchored in reality.

Such unwarranted confidence in outputs can be detrimental, especially in LLM-based robotic planning. This is further exacerbated when natural language instructions, often riddled with inherent or unintentional ambiguities, are provided in real-world settings. Misinterpreting such instructions can lead to undesirable or, in extreme cases, unsafe actions.

To illustrate, the paper presents an example where a robot tasked with heating food is instructed to place a bowl in the microwave. In scenarios where multiple bowls are present, such an instruction becomes ambiguous. Moreover, if one of the bowls is metallic, placing it in the microwave would be hazardous. Instead of acting on such vague instructions, the ideal robot should recognize its uncertainty and seek clarification. While previous works in language-based planning have either overlooked the need for such clarifications or relied heavily on extensive prompting, this paper introduces **KNOWNO**.

KNOWNO is a framework designed to measure and align the uncertainty of LLM-based planners. It ensures that these planners know their limitations and seek assistance when required. The foundation of KNOWNO is built on the theory of conformal prediction, which offers statistical guarantees on task completion while minimizing the need for human intervention in intricate multi-step planning settings. Experiments across various simulated and real robot setups demonstrate the framework's efficacy. These experiments encompass tasks with diverse modes of ambiguity, ranging from spatial uncertainties to numeric ones and from human preferences to Winograd schemas.

The paper posits KNOWNO as a promising lightweight approach to model uncertainty. It can seamlessly complement and scale with the burgeoning capabilities of foundational models. LLMs can be more reliable by leveraging conformal prediction, especially when precision and safety are paramount.

Summary

In the chapter, we have explored the inherent uncertainty challenges in the NLP domain. Recognizing the pivotal role of NLP models in today's critical systems, the chapter emphasizes the importance of ensuring these models' predictions are trustworthy and reliable. The chapter introduces conformal prediction as a solution to address the miscalibration seen in deep learning models' outputs, offering a means to quantify the confidence of predictions robustly. Throughout this chapter, you gained insights into the intricacies of uncertainty quantification specific to NLP, the reasons why deep learning models often produce miscalibrated predictions, and various methods of quantifying uncertainty in NLP. Finally, we deeply studied the conformal prediction technique tailored for NLP tasks.

At the end of this chapter, you should have a holistic understanding of the challenges of uncertainty in NLP, the merits and mechanics of conformal prediction, and practical knowledge to apply this technique to NLP problems effectively.

In the next chapter, we will dive deep into the intriguing world of imbalanced data and show how conformal prediction can address existing challenges in handling such scenarios.

Part 4:
Advanced Topics

This part will provide illustrations on how conformal prediction can be used to solve imbalanced data problems, introducing you to various conformal prediction methods that can be used for multi-class classification problems.

This section has the following chapters:

11

Handling Imbalanced Data

This chapter delves into the intriguing world of imbalanced data and how conformal prediction can be a game-changer in handling such scenarios.

Imbalanced datasets are a common challenge in machine learning, often leading to biased predictions and underperforming models. This chapter will equip you with the knowledge and skills to tackle these issues head-on.

We will be introduced to imbalanced data and learn why it poses a significant challenge in machine learning applications. We will then explore various methods traditionally used to address imbalanced data problems.

The highlight of the chapter is the application of conformal prediction to imbalanced data problems.

This chapter will illustrate how conformal prediction can solve imbalanced data problems by covering the following topics:

- Introducing imbalanced data
- Why imbalanced data problems are complex to solve
- Methods for solving imbalanced data
- How conformal prediction can be applied to help solve imbalanced data problems

Join us on this enlightening journey as we unravel the complexities of imbalanced data and discover innovative solutions through conformal prediction.

By the end of this chapter, you will have a solid understanding of how conformal prediction can be effectively applied to handle imbalanced data, thereby improving the performance and reliability of your machine learning models.

Introducing imbalanced data

In machine learning, we often come across datasets that need to be more balanced. But what does it mean for a dataset to be imbalanced?

An imbalanced dataset is one where the distribution of samples across the different classes is not uniform. In other words, one type has significantly more samples than the other(s). This is a common scenario in many real-world applications. For instance, in a dataset for fraud detection, the number of non-fraudulent transactions (majority class) is typically much higher than the number of fraudulent ones (minority class).

Imagine a medical dataset recording instances of a rare disease. Most patients will be disease-free, resulting in a large class of healthy records, while only a tiny fraction will be affected by the disease. This disproportion in the distribution of categories is what we call imbalanced data.

Imbalanced data can lead to a significant challenge in predictive modeling. By their very nature, machine learning algorithms are designed to minimize errors and maximize accuracy. When trained on imbalanced data, they tend to be biased toward the majority class, often at the expense of the minority class prediction accuracy.

In our medical example, a naive model might predict that no one has the disease, achieving a high accuracy due to the sheer number of healthy records but failing to identify the few crucial cases that do. Such models, misled by the imbalance, could have dire real-world implications.

The nature of imbalanced data is pervasive across industries. From fraud detection in finance, where fraudulent transactions are rare but crucial to detect, to natural disaster predictions in meteorology, where the event of interest (e.g., a tornado or earthquake) is infrequent but significant, imbalances pose challenges that professionals must be equipped to handle.

Recognizing and understanding imbalanced data is the first step in effectively addressing their challenges. As we proceed, we'll deep dive into why these problems are particularly tough to crack and explore methodologies to handle them, focusing on the potential of conformal prediction.

Why imbalanced data problems are complex to solve

Addressing imbalanced data is no walk in the park, and here's why. At the core of the challenge is the nature of conventional machine learning algorithms. These algorithms minimize overall error and are designed with the assumption of balanced class distributions. This becomes problematic when faced with imbalanced datasets, leading to a pronounced bias toward the majority class.

The gravity of this problem becomes evident when we realize that in many scenarios, it's the minority class that carries more significance. Take fraud detection or medical diagnoses as cases in point. While fraudulent transactions or disease instances might be sparse, their correct identification is paramount. Yet, a model trained on skewed data might often lean toward predicting the majority class, achieving superficially high accuracy but failing its core objective.

To add to the challenge, conventional metrics, such as accuracy, are only sometimes our friends here. A dataset with just 2% fraudulent transactions can trick us into complacency: a naive model predicting every transaction as legitimate will boast a 98% accuracy, masking its utter failure in detecting fraud.

The maze of academic literature on this topic makes things even more difficult. With many methods and theories, determining which ones genuinely work is akin to finding a needle in a haystack. Methods such as **Synthetic Minority Oversampling Technique** (**SMOTE**), which is frequently discussed, require discerning analysis to gauge their actual effectiveness.

A word of advice for those just starting in data science: approach the realm of imbalanced classification with a discerning eye. Not all that glitters is gold. While searching for a magic solution is tempting, sometimes it's about reframing the problem. By shifting our perspective and focusing on more relevant metrics, we can find a way through the maze, making informed and effective decisions.

We will now look into some common methods for dealing with imbalanced data.

Methods for solving imbalanced data

Where should we turn when confronted with the challenge of imbalanced class distribution? While a significant portion of resources in the field suggest using resampling methods, including undersampling, oversampling, and techniques such as SMOTE, it's crucial to note that these recommendations often sidestep foundational theory and practical application.

Before diving into solutions for imbalanced classes, it's essential first to understand their underlying nature. The issue might be better approached in specific scenarios such as anomaly detection rather than in a traditional classification problem.

In specific scenarios, the class imbalance isn't static. It can evolve or may be influenced by the need for adequate labels. For instance, consider a system monitoring network traffic for potential security threats. Initially, threats might be rare, leading to a class imbalance. However, as the system matures and more potential hazards are identified and labeled, the imbalance might shift, reducing or reversing the skew.

Addressing such dynamic imbalances requires adaptive methods that can recalibrate as data characteristics change, ensuring the model remains effective throughout its life cycle.

When these challenges are absent, it's prudent to shift focus to the evaluation metrics. We've previously examined metrics such as log loss and Brier loss, which are instrumental in assessing model calibration. Notably, employing resampling techniques with these metrics might adversely impact the model's calibration.

One frequently proposed remedy for imbalanced data is to modify the dataset through various resampling techniques.

Resampling methods are techniques used to balance the distribution of classes in an imbalanced dataset. These methods can be broadly categorized into two main types:

- **Oversampling**: This involves increasing the number of instances in the minority class. Methods include the following:

  - **Random oversampling**: This involves duplicating random records from the minority class.

  - **SMOTE**: SMOTE creates synthetic samples for the minority class in a feature space by following a specific algorithm. It starts by randomly selecting a minority class instance and finding its k-nearest minority class neighbors. SMOTE randomly picks one from these neighbors and calculates the difference between its features and the selected instance's features. It then multiplies this difference by a random number between 0 and 1, adding the result to the original instance's features. This procedure generates a new, synthetic data point that lies somewhere on the line segment, connecting the actual instance with its chosen neighbor, effectively creating plausible new instances that contribute to a more balanced dataset for the classifier to learn from.

  - **Adaptive synthetic (ADASYN) sampling**: Creating synthetic instances for the minority class by following their density distributions. Extra synthetic data is produced for minority samples that pose more significant learning challenges than those that are easier to learn.

- **Undersampling**: This involves reducing the number of instances in the majority class. Methods include the following:

  - **Random undersampling**: This involves randomly eliminating majority class instances.

  - **Tomek links**: This identifies pairs of instances from nearest neighbor classes and removes the majority instance from the pair.

  - **Cluster centroids**: This method replaces a cluster of majority samples with the cluster centroid of a k-means algorithm.

  - **Neighborhood cleaning rule**: This combines undersampling and the **edited nearest neighbor (ENN)** method to remove majority class instances that are misclassified by the KNN classifier and the instances from the minority class that are misclassified.

- **Combining oversampling and undersampling**: Techniques can be used to both oversample the minority class and undersample the majority class to achieve a balance.

- **Ensemble resampling**: This involves creating multiple balanced subsets through resampling and building an ensemble of models.

While resampling methods can help balance the class distribution, they may not always improve model performance, especially in terms of calibration. Evaluating models on a separate, untouched validation set and considering other strategies such as choosing appropriate evaluation metrics is crucial.

While resampling methods such as SMOTE have been accepted for many years as potential solutions, there is no evidence that such methods work across a wide range of datasets. For example, in Kaggle competitions, SMOTE was never successfully used as part of winning solutions.

Over the years, resampling methods, notably SMOTE, have been championed as potential solutions to the challenge of imbalanced datasets. However, a deeper dive into their effectiveness paints a more nuanced picture. Despite their widespread mention in literature and tutorials, there's a conspicuous absence of empirical evidence supporting their efficacy across diverse datasets. A testament to this is the world of Kaggle competitions, where precision, innovation, and effectiveness are paramount. Notably, SMOTE and similar strategies have rarely, if ever, been components of winning solutions. This isn't just a statistical anomaly or coincidence. It underscores a profound observation: while these methods might offer superficial relief in some contexts, they aren't universally applicable or reliably effective. Any practitioner aiming for cutting-edge performance would do well to approach resampling methods with a healthy dose of skepticism and thorough validation.

The study *The harm of class imbalance corrections for risk prediction models: illustration and simulation using logistic regression* by Ruben Van Den Goorbergh, Maarten van Smeden, Dirk Timmerman, Ben Van Calster, investigates the impact of class imbalance adjustments on the performance of logistic regression models. The research scrutinizes conventional and ridge-penalized versions of the model, assessing how these corrections influence their discrimination ability, calibration accuracy, and classification effectiveness.

The paper analyzed techniques such as random undersampling and SMOTE, leveraging both Monte Carlo simulations and a real-world case study on ovarian cancer diagnosis.

Interestingly, while these corrective methods consistently resulted in miscalibrated models (with a pronounced overestimation of the likelihood of falling into the minority class), they didn't necessarily enhance discrimination as measured by the area under the receiver operating characteristic curve. However, they did improve classification metrics such as sensitivity and specificity. Similar classification outcomes could be achieved simply by adjusting the probability threshold.

The paper argues that class imbalance correction techniques can harm the performance of prediction models, particularly in terms of calibration. The research determined that an imbalance in outcomes does not necessarily pose an issue and that attempts to correct this imbalance could degrade the model's performance.

The paper's findings underscore that class imbalance, in isolation, isn't inherently problematic and that efforts to rectify it might inadvertently degrade model performance.

In data science, distinguishing between prediction and classification is pivotal. Classification often mandates a premature decision, merging prediction with the decision-making process, potentially sidelining the actual decision-makers' considerations. This is especially true when the cost of incorrect decisions shifts or data sampling criteria change. On the other hand, predictions remain neutral, serving as tools for any decision-maker.

In his article *Classification vs. Prediction* (`https://www.fharrell.com/post/classification/`), Frank Harell argues that classification can lead to hasty decisions, and its application in machine learning is sometimes misguided. On the other hand, probability modeling quantifies underlying patterns, typically aligning more closely with the core objectives of a project.

Classification is most apt when outcomes are clear-cut, and predictors offer near-certain outcomes. However, many machine learning enthusiasts lean toward classifiers, neglecting the richness of probabilistic thinking, which is deeply rooted in statistics. An example of this is the frequent misclassification of logistic regression as a mere classification tool when, in essence, it offers rich probability estimates.

It's a misconception that binary decisions necessitate binary classifications. Often, the decision might be to gather more data or to take a phased approach. For instance, a physician might opt for progressive treatment based on evolving symptoms rather than making a binary decision upfront.

Consider a high-clarity scenario such as optical character recognition. Here, the outcome is primarily deterministic, and machine learning classifiers excel. However, probability estimates become crucial when there's inherent variability, such as in predicting disease outcomes. They inherently provide error margins, aiding decision-makers in understanding the associated risks.

There's also a challenge with classifiers in imbalanced scenarios. For instance, in a dataset with an overwhelming majority of non-diseased patients, a naive classifier might label everyone as non-diseased, achieving high accuracy but failing in actual detection. Addressing this imbalance often involves practices such as subsampling, which can lead to more issues. Logistic regression, in contrast, can gracefully handle such situations by recalibrating for different datasets or prevalences.

The choice of accuracy metrics is also fundamental. Opting for simplistic accuracy measures can lead to misleading models. The focus should instead be on more nuanced and statistically sound accuracy scoring rules.

In conclusion, while classifiers might be suitable for deterministic scenarios with high-clarity outcomes, for most real-world situations with inherent variability and nuances, probability-based models, such as logistic regression, are more apt, versatile, and insightful.

The issue with resampling methods is that they destroy calibration, which is critical for decision-making; the resampling techniques do not add any new information. The general acceptance of the SMOTE paper that has received over 25K citations is very unfortunate, especially considering that the paper is 20 years old, used only a few datasets, and performed experiments using weak classifiers such as C4.5 (decision tree classifier), Ripper (rule-based algorithm), and a naïve Bayes classifier.

The paper also concentrated on inappropriate metrics, focusing solely on the **area under the curve (AUC)** and the ROC convex hull without considering metrics that measure classifier calibration. Consequently, the paper failed to report the adverse effects on calibration caused by SMOTE.

In the following section, we'll examine effective strategies to address the challenges of imbalanced datasets in machine learning.

The methods for solving imbalanced data

Addressing the challenge of imbalanced data isn't just about achieving a balanced class distribution; it's about understanding the nuances of the problem and adopting a holistic approach that encompasses all facets of model performance. Let us go through the methods for it:

- **Understanding the problem**: The first step is a deep understanding of the problem. It's essential to discern why the data is imbalanced. Is it because of the nature of the data or perhaps due to some external factors or biases in data collection? Recognizing the root cause can offer insights into the most effective strategies.

- **Prioritizing calibration**: One critical aspect that's often overlooked is calibration. A model's ability to provide probability estimates that reflect true likelihoods is paramount, especially when decisions are based on these probabilities. Ensuring the model is well calibrated is often more crucial than mere class separation.

- **Metrics beyond ROC AUC**: While the **receiver operating characteristic area under the curve (ROC AUC)** is a popular metric, relying solely on it can be misleading, especially in imbalanced datasets. It's pivotal to incorporate metrics that capture the essence of calibration. Metrics such as **expected calibration error (ECE)**, log loss, and Brier score, which we've looked into in previous chapters, provide a more comprehensive understanding of a model's performance.

- **Resampling techniques**: While techniques such as oversampling, undersampling, and SMOTE have been propagated as potential solutions, it's crucial to understand their implications. While they might balance class distributions, they may not always improve or even maintain a model's calibration. Therefore, any resampling should be performed cautiously, and the resulting models should be rigorously evaluated on untouched validation sets.

- **Cost-sensitive learning**: Another approach is to assign different costs to misclassifications of the minority and majority classes. By doing so, the algorithm inherently gives more weight to the minority class during training, aiming to reduce the more costly errors.

- **Threshold tuning**: By adjusting the decision threshold away from the default (usually 0.5 for binary classification), one can perform better in the minority class. It's about finding a balance between precision and recall, and this technique can be particularly effective when the real-world costs of false positives and false negatives are different.

Ultimately, the goal is to build effective models differentiating classes and offering calibrated reliable probability estimates. A multifaceted approach emphasizing understanding, calibration, and the right metrics is the way to tackle the imbalanced data problem.

Next, we'll explore how conformal prediction can be applied to help solve imbalanced data problem and offer insights into its potential to enhance data analysis.

Solving imbalanced data problems by applying conformal prediction

Conformal prediction is a technique that can be applied to handle imbalanced data problems. Here are a few ways it can be used:

- **Graceful handling of imbalanced datasets**: conformal prediction can gracefully handle large imbalanced datasets. It strictly defines the level of similarity needed, removing any ambiguity. It can handle severely imbalanced datasets with ratios of 1:100 to 1:1000 without oversampling or undersampling.

- **Local clustering conformal prediction (LCCP)**: LCCP incorporates a dual-layer partitioning approach within the conformal prediction framework. Initially, it segments the imbalanced training dataset into subsets based on class taxonomy. Then, it further divides the examples from the majority class into subsets using clustering techniques. The goal of LCCP is to offer reliable confidence levels for its predictions while also enhancing the efficiency of the prediction process.

- **Mondrian conformal prediction (MCP)**: This can deal with imbalanced datasets. It categorizes data based on their respective labels and assigns a distinct significance level to each class, ensuring that predictive validity is maintained across different classes.

- **Non-conformity scoring**: The core of conformal prediction is the non-conformity measure, which ranks new observations based on how "strange" they appear compared to the training data. This measure can be adapted for imbalanced datasets to give more weight to the minority class, ensuring that the model is more sensitive to the patterns associated with this class.

- **Calibration with validity**: conformal prediction guarantees that if we claim a prediction interval with a 95% confidence level, it will contain the actual outcome 95% of the time in the long run. This built-in calibration, maintained even for imbalanced datasets, ensures that the prediction intervals or sets genuinely reflect the model's uncertainty.

- **Flexibility with underlying models**: conformal prediction is not tied to a specific machine learning algorithm. This means that, even in the context of imbalanced data, practitioners can choose the best-performing base model (a tree-based method, neural network, or linear model) and then apply the conformal framework to obtain reliable predictions.

- **Transparency and interpretability**: The conformal prediction framework's transparent nature allows straightforward interpretation. This transparency can be invaluable for imbalanced datasets, enabling stakeholders to understand why specific predictions are made and how certain the model is about them.

- **Adaptive to changing distributions**: One of the challenges with imbalanced data is that the distribution of the minority class can change over time. With its emphasis on ranking new observations based on their non-conformity, conformal prediction can adapt to these changes, ensuring that predictions remain calibrated even as the underlying data distribution evolves.

conformal prediction provides a framework that can be adapted to handle imbalanced datasets in various ways, offering potential solutions to this common problem in machine learning. While classification is now commonplace, the ultimate goal is enabling informed decisions, which requires reliable probability estimates even with skewed class data.

Addressing imbalanced data with Venn-Abers predictors

In the ever-evolving world of machine learning, addressing classification problems has become commonplace. From distinguishing between cats and dogs to more intricate challenges, the real aim of classification isn't merely labeling; it's also facilitating informed decision-making. For this purpose, more than just class labels are required. We need well-calibrated class probabilities.

Most data scientists, especially those in the early stages of their journey, tend to evaluate classification models using standard metrics such as accuracy, precision, and recall. While these metrics are insightful for more straightforward tasks, they can be misleading for more intricate, real-world problems. The true essence of classification lies in calibration, an aspect often overlooked in introductory courses.

For professionals working on critical applications, from finance to healthcare, the calibration of classifiers is paramount. The heart of a classification problem is to make informed decisions. These decisions revolve around the probabilities of various scenarios, each with potential costs and benefits.

Take the banking sector, for instance. If a model merely predicts that a potential customer won't default on a loan, it needs to provide more depth for decision-making, especially when substantial amounts of money are at stake. What's needed is a model that offers well-calibrated probabilities of various outcomes, allowing for a nuanced evaluation of risks and rewards.

However, a significant challenge arises: many machine learning models don't inherently produce class probabilities. Even if they do, these probabilities can often be miscalibrated, leading to erroneous decision-making. This is particularly concerning in critical sectors. For example, a self-driving car that misinterprets an obstacle due to miscalibrated probabilities could result in accidents.

So, what can be done to achieve better calibration? Classic methods, such as Platt's scaling (`https://en.wikipedia.org/wiki/Platt_scaling`) and isotonic regression (`https://en.wikipedia.org/wiki/Isotonic_regression`), were early solutions. However, these methods have limitations, often rooted in restrictive assumptions that hamper their efficacy across diverse datasets.

Enter **Venn-Abers predictors**, a beacon of hope in classifier calibration. Venn-Abers predictors, a subset of the conformal prediction framework, promise a more robust approach to calibration. Unlike traditional methods, they don't hinge on overly simplistic assumptions and offer a more versatile calibration tool apt for today's complex datasets.

In essence, if you aim to harness the true potential of machine learning classifiers in 2022 and beyond, Venn-Abers and the broader conformal prediction framework are worth exploring. They might be the key to unlocking well-calibrated, reliable machine learning models.

Venn-Abers predictors stand out in machine learning, offering probability-driven predictions for test data labels. What sets them apart is their built-in assurance of calibration. This assurance is grounded in the typical premise that data observations are independently sourced from a consistent distribution.

At its core, the Venn-Abers approach is inspired by isotonic regression. It refines the probabilistic prediction calibration method pioneered by Zadrozny and Elkan. In contrast to techniques such as Platt's scaler and isotonic regression, Venn-Abers predictors come equipped with inherent mathematical proofs, ensuring their unbiased validity.

An intriguing feature of Venn-Abers predictors is their ability to produce dual probability predictions for the *class 1* label. This dual output captures the range of prediction uncertainty. As a result, these predictors offer calibrated predictions and shed light on the inherent confidence associated with each prediction. This makes them invaluable tools for enhancing the calibration of probability-based predictions. Here's how:

- **True-to-life probability intervals**: Venn-Abers predictors shine in delivering calibrated probability intervals. This ensures that the probabilities they produce genuinely represent the actual chances of an event, eliminating the pitfalls of overconfidence or underestimation.

- **Versatility across models**: The beauty of Venn-Abers calibration is its adaptability. Whether you're working with decision trees, random forests, or even XGBoost models, Venn-Abers can recalibrate them, fine-tuning overambitious and cautious models to enhance their accuracy.

- **Enhanced decision support with valid intervals**: The predictors don't just stop at labels. For every prediction, especially from typically complex models such as random forests and XGBoost, Venn-Abers offers a probability interval. The span of this interval serves as a barometer of the prediction's reliability

Venn-Abers predictors are a beacon for those navigating the choppy waters of imbalanced data issues. They refine the predictive accuracy of various machine learning models and arm users with credible probability intervals, making decision-making more informed and confident.

To illustrate the various issues in imbalanced data problems, we will use the following notebook: `https://github.com/PacktPublishing/Practical-Guide-to-Applied-Conformal-Prediction/blob/main/Chapter_11.ipynb`

This notebook will look at various methods for handling an imbalanced class problem and apply conformal prediction to calibrate class probabilities.

We will use the Credit Card Fraud Detection dataset from Kaggle: `https://www.kaggle.com/datasets/mlg-ulb/creditcardfraud`

The dataset contains data on credit card transactions in September 2013 by cardholders in Europe. The transactions occurred over two days, with 492 fraudulent transactions out of 284,807 transactions. The dataset is highly imbalanced, with the positive class (fraudulent transactions) accounting for 0.17% of all transactions.

The dataset contains numerical features that are the result of PCA transformation; the original features have been withheld due to confidentiality and privacy issues.

Features V1, V2, ... V28 are the principal components obtained using PCA:

- The only original features are Time and Amount

- The feature Time contains the time (in seconds) for each transaction relative to the first transaction in the dataset

- The feature Amount is the transaction amount

- The label Class is the dependent variable that needs to be predicted (fraudulent transactions labeled with 1)

We will use various classifiers, including popular classifiers such as XGBoost, LightGBM, CatBoost, Random Forest, and logistic regression.

Key insights from the Credit Card Fraud Detection notebook

In our exploration of the Credit Card Fraud Detection dataset, we unearthed several pivotal insights that can reshape our approach to imbalanced data:

- **Embracing simplicity**: The most effective strategy is often to leave the data untouched. Contrary to the push for intricate resampling techniques, a minimalist approach can sometimes yield superior results.

- **Reframing imbalance**: Rather than viewing imbalanced data as a dilemma needing a direct fix, it's crucial to understand that the imbalance isn't always the root issue. The quest shouldn't be to balance the scales but to derive meaningful insights from the data, irrespective of distribution.

- **The power of robust metrics**: The choice of metrics can make or break your analysis. By employing a comprehensive set of metrics, you can accurately define the problem and pave the way for practical solutions.

- **Calibration's central role**: Calibration is non-negotiable in real-world decision-making scenarios, especially in critical applications. Accurate probability estimations are vital, ensuring decisions are based on reliable data.

- **The double-edged sword of resampling**: While resampling methods might seem promising, they often compromise the model's calibration. Our analysis demonstrated that such techniques could deteriorate calibration metrics such as ECE, log loss, and Brier score.

- **conformal prediction as a beacon**: Amid the challenges posed by imbalanced data and the potential pitfalls of resampling, conformal prediction emerges as a silver lining. It offers a reliable method to recalibrate probabilities, ensuring that even post-resampling, the data remains conducive for sound decision-making.

By internalizing these insights, we can approach imbalanced datasets with a refined perspective, prioritizing meaningful analysis over superficial fixes.

Summary

The challenge of imbalanced datasets in machine learning often results in biased predictions and compromised model outcomes. This chapter delves deep into the complexities of such datasets and illuminates the path through conformal prediction, a groundbreaking approach to handling these scenarios.

Traditional methods, such as resampling techniques, and metrics, such as ROC AUC, often fail to address the imbalances effectively. Furthermore, they can sometimes lead to even more skewed results. On the other hand, conformal prediction emerges as a robust solution, offering calibrated and reliable probability estimates.

The practical implications of these methods are illustrated using the Credit Card Fraud Detection dataset from Kaggle, an inherently imbalanced dataset. The exploration underscores the significance of understanding the data, using robust metrics, and the transformative potential of conformal prediction.

In essence, while imbalanced data presents challenges, practitioners can navigate toward calibrated and insightful predictions with the right tools such as conformal prediction.

In the next chapter of this book, we will dive deep into the fascinating world of multi-class conformal prediction. This chapter will introduce you to various conformal prediction methods that can be effectively applied to multi-class classification problems.

12

Multi-Class Conformal Prediction

Welcome to the last chapter of this book, where we delve into the fascinating world of multi-class **Conformal Prediction**. This chapter introduces you to various conformal prediction methods that can be effectively applied to multi-class classification problems.

We will explore the concept of multi-class classification, a common scenario in **machine learning (ML)**, where an instance can belong to one of many classes. Understanding this problem is the first step toward applying conformal prediction techniques effectively.

Next, we will investigate the metrics used to evaluate multi-class classification problems. These metrics provide a quantitative measure of the performance of our models, and understanding them is crucial for effective model evaluation and selection.

Finally, we will learn how to apply conformal prediction to multi-class classification problems. This section will provide practical insights and techniques to apply directly to your industrial applications.

By the end of this chapter, you will have gained valuable skills and knowledge in multi-class classification and learned how conformal prediction can be effectively applied to these problems. So, let's dive in and start our journey into multi-class conformal prediction!

In this chapter, we're going to cover the following main topics:

- Multi-class classification problems
- Metrics for multi-class classification problems
- How conformal prediction can be applied to multi-class classification problems

Multi-class classification problems

In ML, classification problems are ubiquitous. They involve predicting a discrete class label output for an instance. While binary classification – predicting one of two possible outcomes – is a common scenario, many real-world problems require predicting more than two classes. This is where multi-class classification comes into play.

Multi-class classification is a problem where an instance can belong to one of many classes. For example, consider an ML model designed to categorize news articles into topics. The articles could be classified into categories such as *Sports*, *Politics*, *Technology*, *Health*, and so on. Each of these categories represents a class, and since there are more than two classes, this is a multi-class classification problem.

It's important to note that each instance belongs to exactly one class in multi-class classification. If each instance could belong to multiple classes, it would be a multi-label classification problem, which is a different kind of problem.

Multi-class classification problems are a staple in ML, and they require a slightly different approach than binary classification problems. Let's dive deeper into the intricacies of multi-class classification.

Algorithms for multi-class classification

Several ML algorithms can handle multi-class classification problems directly. These include, but are not limited to, the following:

- **Decision trees**: Decision tree algorithms such as **Classification and Regression Trees** (**CARTs**) can naturally handle multi-class classification.
- **Naive Bayes**: Naive Bayes treats each class as a separate one-versus-all binary classification problem and picks the outcome with the highest probability.
- **Neural networks** (**NNs**): NNs can be designed with an output layer of multiple nodes representing a class. The `softmax` activation function can then be used to calculate the probability distribution of each class.

Many ML algorithms are inherently designed for binary classification between two classes. Special strategies must be employed to extend these models to multi-class problems with more than two categories. Two common approaches are one-vs-all and one-vs-one.

Next, we will explore these one-vs-all and one-vs-one strategies in more detail, including how binary classification outcomes get aggregated to make final multi-class predictions. We will also discuss evaluating multi-class classifiers using specialized performance metrics suited for problems with more than two categories.

One-vs-all and one-vs-one strategies

For algorithms that do not natively support multi-class classification, strategies such as one-vs-all (also known as one-vs-rest) and one-vs-one are used as follows:

- **One-vs-all strategy**: For a problem with n classes, n separate binary classification models are trained. Each model is trained to distinguish instances of one class from instances of all other classes. All n models are applied for a new instance, and the model that gives the highest confidence score determines the instance's class.

- **One-vs-one strategy**: A binary classification model is trained for every pair of classes in this strategy. For n classes, this results in $n(n-1)/2$ models. Each model's decision contributes to a voting scheme, and the class with the most votes is chosen as the final class of the instance. For example, for a 4-class problem, $4*3/2 = 6$ binary models would be built, one for each pair of classes. Each model casts a vote for its predicted class, and the class with the most votes across all models is chosen as the final prediction. So, with four classes, if three models predicted class A, two predicted class B, and one predicted class C, class A would be selected since it received the most votes. In this way, each model's decision contributes to a voting scheme to determine the overall predicted class.

The following section will discuss the metrics used for evaluating multi-class classification problems. Understanding these metrics is crucial for assessing the performance of our models and making informed decisions about model selection and optimization.

Metrics for multi-class classification problems

In the multi-class classification field, evaluating models' performance is as crucial as developing them. Effective evaluation hinges upon utilizing the right metrics that can accurately measure the performance of the multi-class classification models and provide insights for improvement. This section demystifies the various metrics essential for assessing the performance of multi-class classification models, providing a solid foundation for selecting and employing the right metric for your specific use case.

Confusion matrix

One of the fundamental metrics for evaluating multi-class classification models is the **confusion matrix.** It provides a visualization of the performance of an algorithm, typically a **supervised learning (SL)** one. Each row of the confusion matrix represents the instances of an actual class, while each column represents the instances of a predicted class. It's an essential tool for understanding the model's performance beyond overall accuracy, offering insight into classification errors.

Precision

Precision (or **positive predictive value**; **PPV** in short) is a measure that examines the number of true positive predictions among the total positive predictions made by the model. High precision indicates that the false positive rate is low. For multi-class classification problems, precision is calculated for each class separately and can be averaged to understand the overall performance.

Recall

Recall (or **sensitivity** or **true positive rate**; **TPR** in short) gauges the number of true positive predictions among the actual positives. It is a crucial metric for problems where identifying all actual positives is essential. As with precision, recall is calculated for each class and can be averaged for overall performance assessment.

F1 score

F1 score is the harmonic mean of precision and recall, balancing the two metrics. It is particularly useful when dealing with imbalanced datasets, offering a more holistic view of the model's performance beyond accuracy.

Macro- and micro-averaged metrics

In multi-class classification problems, averaging metrics such as precision, recall, and F1 score, commonly known as macro- and micro-averaging, can be done in multiple ways:

- **Macro-averaging** computes the metric independently for each class and then takes the average, treating all classes equally

- **Micro-averaging** aggregates the contributions of all classes to compute the average metric

Area Under Curve (AUC-ROC)

Another important metric is the **Area Under the Receiver Operating Characteristic Curve** (**AUC-ROC**). While it is primarily used for binary classification problems, it can be extended to multi-class classification by considering each class against the rest.

Log loss and its application in measuring calibration of multi-class models

Log loss, also known as **logistic loss** or **cross-entropy loss**, is a commonly used loss function for classification problems, including multi-class classification. It quantifies the performance of a classification model by measuring the uncertainty in the predictions. Log loss assigns a penalty for incorrect classifications; the penalty is higher for confidently wrong predictions.

Mathematical representation

Mathematically, log loss for multi-class classification can be represented as follows:

$$-\frac{1}{N}\sum_{i=1}^{N}\sum_{j=1}^{M}y_{ij}\log(p_{ij})$$

Here, the following apply:

- N is the number of observations
- M is the number of classes.
- y_{ij} is a binary indicator for whether class j is the correct classification for observation i.
- p_{ij} is the predicted probability that observation i is of class j.

Using log loss to measure calibration

A calibrated model is one whose predicted probabilities reliably reflect the true likelihood of the predicted outcomes. Calibration in multi-class classification means that if a model predicts a class with a probability p, then that class should occur about p percent of the time among all instances where that class is predicted with probability p.

Log loss and calibration

Log loss is an appropriate metric to assess the calibration of a multi-class classification model because it directly compares the predicted probabilities (the confidence of the predictions) with the actual classes. A well-calibrated model will have a lower log loss as the predicted probabilities for the actual classes will be higher.

How to evaluate calibration using Log loss

- **Predict the probabilities**: Use your multi-class classification model to predict the probabilities for each class for each observation in a validation dataset
- **Compute log loss**: Calculate the log loss using the formula previously shown
- **Interpret the result**: A lower log loss value indicates better calibration as it shows that the predicted probabilities are closer to the actual classes

Evaluating the log loss of your multi-class classification model gives you insights into the calibration of the model. A model with a lower log loss is more calibrated, providing more reliable prediction probability estimates. Understanding and using log loss as a metric to measure the calibration is vital to ensure that your multi-class classification model performs optimally in real-world applications.

Brier score and its application in measuring the calibration of multi-class models

The Brier score, or quadratic loss, is another popular metric used to evaluate the performance of classification models, including multi-class classification problems. It quantifies the difference between the predicted probabilities and the actual classes, assigning a lower score to better-calibrated models.

Mathematical representation

For multi-class classification, the Brier score is calculated as follows:

$$\frac{1}{N}\sum_{i=1}^{N}\sum_{j=1}^{M}(p_{ij} - o_{ij})^2$$

Here, the following apply:

- N is the number of observations
- M is the number of classes.
- y_{ij} is a binary indicator for whether class j is the correct classification for observation i.
- p_{ij} is the predicted probability that observation i is of class j.

Using the Brier score to measure calibration

The Brier score is an effective metric for assessing the calibration of multi-class models as it penalizes the model more when there is a larger difference between the predicted probabilities and the actual outcomes. A well-calibrated model will have a lower Brier score as its predicted probabilities will be closer to the actual outcomes.

How to evaluate calibration using Brier Score

The Brier score provides a way to quantitatively assess how well calibrated a multi-class classifier's predicted probabilities are. Evaluating calibration is key for ensuring reliability in real-world deployment.

To utilize the Brier score, there are three main steps:

1. **Predict the probabilities**: Use your multi-class classification model to estimate the probabilities for each class for every observation in a validation dataset.
2. **Compute the Brier score**: Calculate the Brier score using the provided formula.
3. **Interpret the result**: A lower Brier score indicates better calibration. It signifies that the model's predicted probabilities are more aligned with the actual outcomes, thus making the model more reliable.

In essence, employing the Brier score to evaluate the calibration of your multi-class model helps ensure the reliability of your model's probability estimates. A lower Brier score, reflecting smaller differences between predicted and actual probabilities, indicates a well-calibrated model, enhancing the model's trustworthiness in real-world applications. Understanding and utilizing the Brier score as a calibration metric is essential for optimizing the performance of your multi-class classification model in practical scenarios.

Understanding and employing the appropriate metrics is pivotal for evaluating and improving multi-class classification models. A thorough grasp of these metrics allows for a more nuanced analysis, paving the way for developing robust and efficient multi-class classification models and ensuring their successful deployment in real-world scenarios.

How conformal prediction can be applied to multi-class classification problems

conformal prediction is a powerful framework that can be applied to multi-class classification problems. It provides a way to make predictions with a measure of certainty, which is particularly useful when dealing with multiple classes.

In the previous chapters, we have already looked at how conformal prediction assigns a p-value to each class for a given instance in the context of multi-class classification.

The p-value represents the confidence level of the prediction for that class. The higher the p-value, the more confident the model is that the instance belongs to that class.

The procedure for applying conformal prediction to multi-class classification is as follows:

1. **Calibration**: A portion of the training data, known as the calibration set, is set aside. The model is trained on the remaining data.

2. **Prediction**: For each class, the model predicts class scores. The conformity score, which measures how well the prediction conforms to the actual outcomes in the calibration set, is calculated.

3. **P-value calculation**: For a new instance, the model calculates a nonconformity score for each class. The p-value for each class is then calculated as the proportion of instances in the calibration set with a higher nonconformity score.

4. **Output**: The model outputs the predicted class labels along with their p-values. The classes are ranked by their p-values, providing a measure of confidence for each prediction.

Applying conformal prediction to multi-class classification problems offers several benefits:

- **Confidence measures**: conformal prediction provides a measure of confidence (the p-value) for each prediction, which can be very useful in decision-making processes
- **Validity**: conformal prediction offers a theoretical guarantee of validity, meaning that the error rate of the predictions will be close to the significance level set by the user
- **Efficiency**: conformal prediction is computationally efficient and can be applied to large datasets
- **Versatility**: conformal prediction can be used with any ML algorithm, making it a versatile tool for multi-class classification problems

The next section will examine how Venn-ABERS predictors can be applied to multi-class classification problems.

Multi-class probabilistic classification using inductive and cross-Venn-ABERS predictors

Venn-ABERS is a conformal prediction method developed by Vladimir Vovk, Ivan Petej, and Valentina Fedorova (*Large-scale probabilistic predictors with and without guarantees of validity*, `https://papers.nips.cc/paper/2015/hash/a9a1d5317a33ae8cef33961c34144f84-Abstract.html`) to address the limitations of classical calibrators such as Platt scaling and isotonic regression. It guarantees mathematical validity, regardless of the data distribution, dataset size, or underlying classification model.

Venn-ABERS predictors work by fitting isotonic regression twice, assuming that each test object can have both the label 0 and the label 1. This results in two probabilities, $p0$ and $p1$, for each test object, representing probabilities of the object belonging to class 1. These probabilities create a prediction interval for the probability of class 1, with mathematical guarantees that the actual probability falls within this interval.

The Venn-ABERS prediction is a multi-predictor, and the width of the interval (p0, p1) contains valuable information about the classification confidence. In critical situations, the Venn-ABERS predictor outputs accurate and well-calibrated probabilities and issues an "alert" by widening the (p0, p1) interval. This alert indicates that the decision-making process should consider the increased uncertainty.

The probabilities can be combined into a single value using $p = p1 / (1 - p0 + p1)$ for practical decision-making purposes. This combined probability of class 1, p, can be used for decision-making tasks.

The research paper by Valery Manokhin, *Multi-class probabilistic classification using inductive and cross Venn–Abers predictors*, introduces a method for adapting Venn-ABERS predictors for multi-class classification. You can access the paper here: `http://proceedings.mlr.press/v60/manokhin17a/manokhin17a.pdf`.

Experimental results demonstrate that the proposed multi-class predictors outperform uncalibrated and existing classical calibration methods in terms of accuracy, indicating potential substantial advancements in multi-class probabilistic classification.

For practitioners and researchers eager to employ this technique, a Python implementation of the fast Venn-ABERS predictor for binary classification is accessible on GitHub (*Multi-class probabilistic classification using inductive and cross Venn–Abers predictors*: https://github.com/valeman/Multi-class-probabilistic-classification). This educational resource offers a hands-on opportunity to explore the details of implementation and practical advantages of utilizing Venn-ABERS predictors in real-world ML scenarios.

The proposed approach to multi-class probability estimation using **inductive** and **cross-Venn-ABERS predictors** (**IVAPs** and **CVAPs**) is based on transforming multi-class classifiers into binary classifiers. In this approach, the binary classifiers are trained to distinguish between each class and all other classes combined.

For example, in a three-class problem with classes A, B, and C, three binary classifiers are trained: one to distinguish between A and (B or C), one to distinguish between B and (A or C), and one to distinguish between C and (A or B). The IVAP is then used to estimate the probability of each class for a given test instance. The IVAP computes the probability of a class as the fraction of binary classifiers that classify the instance as belonging to that class.

The formula for converting pairwise classification scores and pairwise class probabilities uses a method introduced in the paper *Pairwise Neural Network Classifiers with Probabilistic Outputs* (https://proceedings.neurips.cc/paper_files/paper/1994/file/210f760a89db30aa72ca258a3483cc7f-Paper.pdf).

Specifically, if r_{ij} is the class score i over class j from the respective binary model, the estimated probability for class i is calculated as follows.

The key idea is to first compute pairwise probabilities between each pair of classes using the dedicated binary classifiers. Then, these pairwise probabilities can be combined to estimate a normalized probability distribution over all classes:

$$p_i^{\text{PKPD}} = \frac{1}{\sum_{j:j\neq i}^{n} \frac{1}{r_{ij}} - (k-2)}$$

This provides a principled approach to converting pairwise binary classification outcomes into class probability estimates suitable for multi-class evaluation and calibration.

After calculating the probabilities (this technique is dubbed the **PKPD** method), normalizing these values is crucial to ensure their total sum is 1. These pairwise probabilities can be obtained by applying IVAPs and CVAPs to the pairwise classification scores/probabilities, which are used to calibrate classification scores produced by underlying classification models.

In simpler terms, the PKPD method helps convert probabilities from binary comparisons (pairwise probabilities) into multi-class probabilities. The calculated multi-class probabilities are then used to classify test objects into one of the k possible classes. This classification enables the computation of metrics, which can be compared across various calibration algorithms to evaluate their performance.

Now, let's explore a real-world example to understand how to apply conformal prediction to multi-class classification problems. The following is a code walk-through that demonstrates this application.

Let's analyze the code from the `Chapter_12.ipynb` notebook (the code can be found in the book's GitHub repo at `https://github.com/PacktPublishing/Practical-Guide-to-Applied-Conformal-Prediction/blob/main/Chapter_12.ipynb`). The notebook demonstrates the application of conformal prediction to a multi-class classification problem.

The general steps of multi-class classification are as follows:

1. **Data preparation**:

   I.   Load the dataset.

   II.   Divide the dataset into features (X) and target (y).

   III. Split the data into training and test sets.

2. **Model training**:

   I.   Train a classification model on the training data.

   II.   Use the trained model to make predictions on the test data.

3. **Applying conformal prediction**:

   I.   Apply conformal prediction to the trained model.

   II.   Obtain confidence and credibility measures for the predictions.

4. **Evaluation**:

   I.   Evaluate the model using appropriate metrics.

   II.   Compare the performance of the calibrated model with the original model.

5. **Model evaluation function**:

   I.   The `evaluate_model_performance` function is used to evaluate the performance of different models and calibration methods.

   II.   It trains the model, makes predictions, and evaluates the performance using various metrics such as accuracy, log loss, and Brier loss.

6. **Calibration:**

   I.   Different calibration methods, such as Platt, isotonic, and others, are applied to the model.

   II.   The predictions of the calibrated models are evaluated to analyze the performance improvement.

7. **Model processing loop:**

   I.   A loop processes different models and applies the `evaluate_model_performance` function to each model.

   II.   The results for each model and calibration method are stored and can be analyzed to determine the best-performing model and calibration method.

Let's summarize the chapter next.

Summary

In this book's final chapter, we explored the intriguing domain of multi-class conformal prediction. We began by understanding the concept of multi-class classification, a prevalent scenario in ML where an instance can belong to one of many classes. This understanding is crucial for effectively applying conformal prediction techniques.

We then delved into the metrics used for evaluating multi-class classification problems. These metrics quantitatively measure our model's performance and are vital for effective model evaluation and selection.

Finally, we learned how to apply conformal prediction to multi-class classification problems. This section provided practical insights and techniques to apply to your industrial applications directly.

By the end of this chapter, you should have gained valuable skills and knowledge in multi-class classification and how conformal prediction can be effectively applied to these problems. This knowledge will prove invaluable in your journey as a data scientist, ML engineer, or researcher.

We covered different main topics, including multi-class classification problems, where we explored multi-class classification and its importance in ML. We also discussed the difference between multi-class and multi-label classification problems. We then investigated the metrics used to evaluate multi-class classification problems. Understanding these metrics is crucial for assessing the performance of our models and making informed decisions about model selection and optimization. Finally, we learned how to apply conformal prediction to multi-class classification problems. This section provided practical insights and techniques to apply directly to your industrial applications.

This chapter marks the end of our journey into conformal prediction. We hope that the knowledge and skills you've gained will serve you well in your future endeavors in ML. Happy learning!

Index

‹packt›

Subscribe to our online digital library for full access to over 7,000 books and videos, as well as industry leading tools to help you plan your personal development and advance your career. For more information, please visit our website.

Why subscribe?

- Spend less time learning and more time coding with practical eBooks and Videos from over 4,000 industry professionals

- Improve your learning with Skill Plans built especially for you

- Get a free eBook or video every month

- Fully searchable for easy access to vital information

- Copy and paste, print, and bookmark content

Did you know that Packt offers eBook versions of every book published, with PDF and ePub files available? You can upgrade to the eBook version at packtpub.com and as a print book customer, you are entitled to a discount on the eBook copy. Get in touch with us at customercare@packtpub.com for more details.

At www.packtpub.com, you can also read a collection of free technical articles, sign up for a range of free newsletters, and receive exclusive discounts and offers on Packt books and eBooks.

Other Books You May Enjoy

If you enjoyed this book, you may be interested in these other books by Packt:

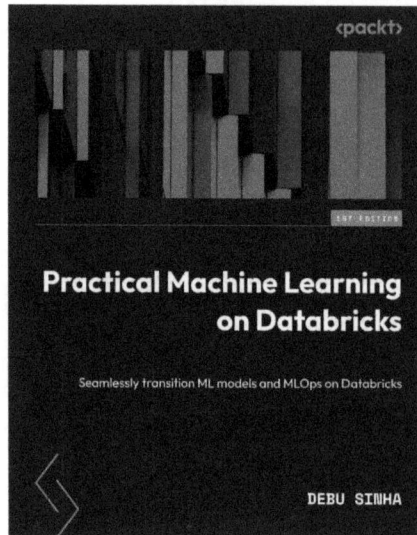

Practical Machine Learning on Databricks

Debu Sinha

ISBN: 978-1-80181-203-0

- Transition smoothly from DIY setups to databricks
- Master AutoML for quick ML experiment setup
- Automate model retraining and deployment
- Leverage databricks feature store for data prep
- Use MLflow for effective experiment tracking
- Gain practical insights for scalable ML solutions
- Find out how to handle model drifts in production environments

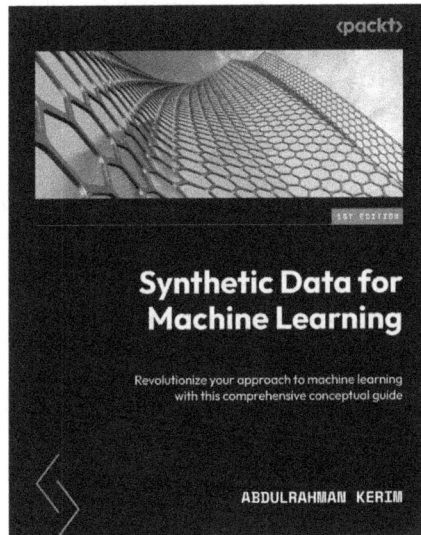

Synthetic Data for Machine Learning

Abdulrahman Kerim

ISBN: 978-1-80324-540-9

- Understand real data problems, limitations, drawbacks, and pitfalls
- Harness the potential of synthetic data for data-hungry ML models
- Discover state-of-the-art synthetic data generation approaches and solutions
- Uncover synthetic data potential by working on diverse case studies
- Understand synthetic data challenges and emerging research topics
- Apply synthetic data to your ML projects successfully

Packt is searching for authors like you

If you're interested in becoming an author for Packt, please visit `authors.packtpub.com` and apply today. We have worked with thousands of developers and tech professionals, just like you, to help them share their insight with the global tech community. You can make a general application, apply for a specific hot topic that we are recruiting an author for, or submit your own idea.

Share Your Thoughts

Now you've finished *Practical Guide to Applied Conformal Prediction in Python*, we'd love to hear your thoughts! If you purchased the book from Amazon, please click here to go straight to the Amazon review page for this book and share your feedback or leave a review on the site that you purchased it from.

Your review is important to us and the tech community and will help us make sure we're delivering excellent quality content.

Download a free PDF copy of this book

Thanks for purchasing this book!

Do you like to read on the go but are unable to carry your print books everywhere? Is your eBook purchase not compatible with the device of your choice?

Don't worry, now with every Packt book you get a DRM-free PDF version of that book at no cost.

Read anywhere, any place, on any device. Search, copy, and paste code from your favorite technical books directly into your application.

The perks don't stop there, you can get exclusive access to discounts, newsletters, and great free content in your inbox daily

Follow these simple steps to get the benefits:

1. Scan the QR code or visit the link below

https://packt.link/free-ebook/9781805122760

2. Submit your proof of purchase
3. That's it! We'll send your free PDF and other benefits to your email directly

www.ingramcontent.com/pod-product-compliance
Lightning Source LLC
Chambersburg PA
CBHW081101220326
41598CB00038B/7187